DARWIN'S SAVAGES

MATTHEW CARR

Darwin's Savages

Science, Race and the Conquest of Patagonia

HURST & COMPANY, LONDON

First published in the United Kingdom in 2025 by
C. Hurst & Co. (Publishers) Ltd.,
New Wing, Somerset House, Strand, London WC2R 1LA

Distributed in the United States, Canada and Latin America by Oxford
University Press, 198 Madison Avenue, New York, NY 10016, United
States of America.

A Cataloguing-in-Publication data record for this book
is available from the British Library.

ISBN: 9781805262831

Printed and bound in Great Britain by Bell & Bain Ltd, Glasgow

www.hurstpublishers.com

CONTENTS

**Part III
Conquest**

Pacific Ocean

Río Desaguadero

Córdoba

Santa Fe

URUGUAY

Mendoza

San Luis

Valparaíso

Río Paraná

Buenos Aires

Isla Martín García

Santiago

ARGENTINA

Junín

Montevideo

La Plata

Río de la Plata

Veinticinco de Mayo

Andes Mts.

Trenque Lauquen

Río Salado

Tapalqué

Concepción

Leubucó

Azul

Carhué

Río Bío bío

Río Colorado

Sierra de la Ventana

Tandil

CHILE

Río Neuquén

Bahía Blanca

Tres Arroyos

Neuquén ☆

Choele Choel

Punta Alta

Valdivia

Río Limay

Río Negro

Valcheta

Viedma

Carmen de Patagones (Fort Carmen)

Puerto Montt

Bariloche

Isla de Chiloé

Esquel

Puerto Madryn

Tecka

Trelew

Trevelin

Río Chubut

Rawson

Atlantic Ocean

Andes Mts.

Río Deseado

PATAGONIA

Puerto Deseado

Falklands / Malvinas Islands

Puerto San Julián

Río Santa Cruz

El Calafate

Puerto Natales

Bahía San Gregorio

Punta Arenas

Bahía Inútil

Tierra del Fuego (Isla Grande)

Río Grande

Strait of Magellan

Beagle Channel

Isla de los Estados

Isla Dawson

Ushuaia

Le Maire Strait

Wulaia Cove

Puerto Williams

Isla Navarino

Cape Horn

Islas Wollaston

N

0   120
miles

☆ Settlements founded in the 20th Century

© S.Ballard (2025)

Map 1: Argentina and Chile

CHILE

5 July 1835

Copiapó

Huasco

Coquimbo

Illapel

Quillota
Valparaíso

Nov 1834

Mendoza

Santiago

Curico

Concepción

Valdivia

Isla de Chiloé

28 Jun –
13 Jul 1834

Dec 1834
– Jan 1835

Río Salado

ARGENTINA

Río Paraná

Río Uruguay

Corrientes

URUGUAY

Santa Fé

27 Sept – 21 Oct 1833

Rosario

Mercedes
6-28 Nov 1833

Minas

Buenos Aires

Montevideo

Río Desaguadero

Río Atuel

Tapalqué

Río Salado

Río Colorado

Bahía
Blanca

8-20 Sept
1833

Río Neuquén

Río Negro

4-17 Aug
1833

Río Limay

Carmen de Patagones
(Fort Carmen)

Apr
1833

Dec 1833

Nov
1832

Río Chubut

PATAGONIA

Río Deseado

Río Chico

Puerto Deseado

Puerto San Julián

5-13 April 1833

Falklands /
Malvinas Islands

Tierra
del Fuego

1 Mar –
5 Apr 1833

10 Mar –
7 Apr 1834

2 Dec 1832 - 26 Feb 1833
Dec 1833 - 5 May 1834
(Beagle surveying around Tierra del Fuego)

N

0   150
miles

© S.Ballard (2025)

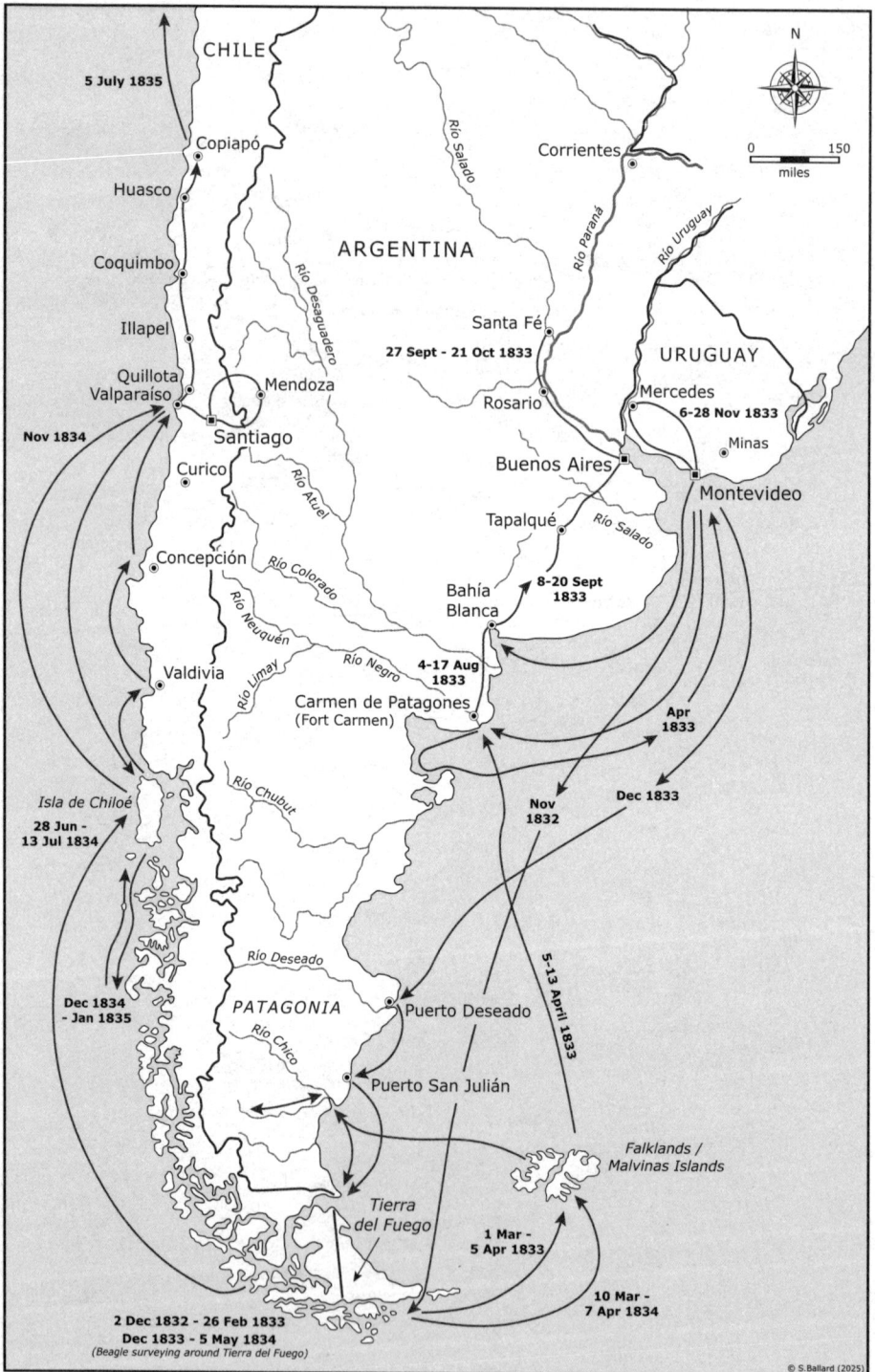

Map 2: Darwin's travels around South America between 1832 and 1835

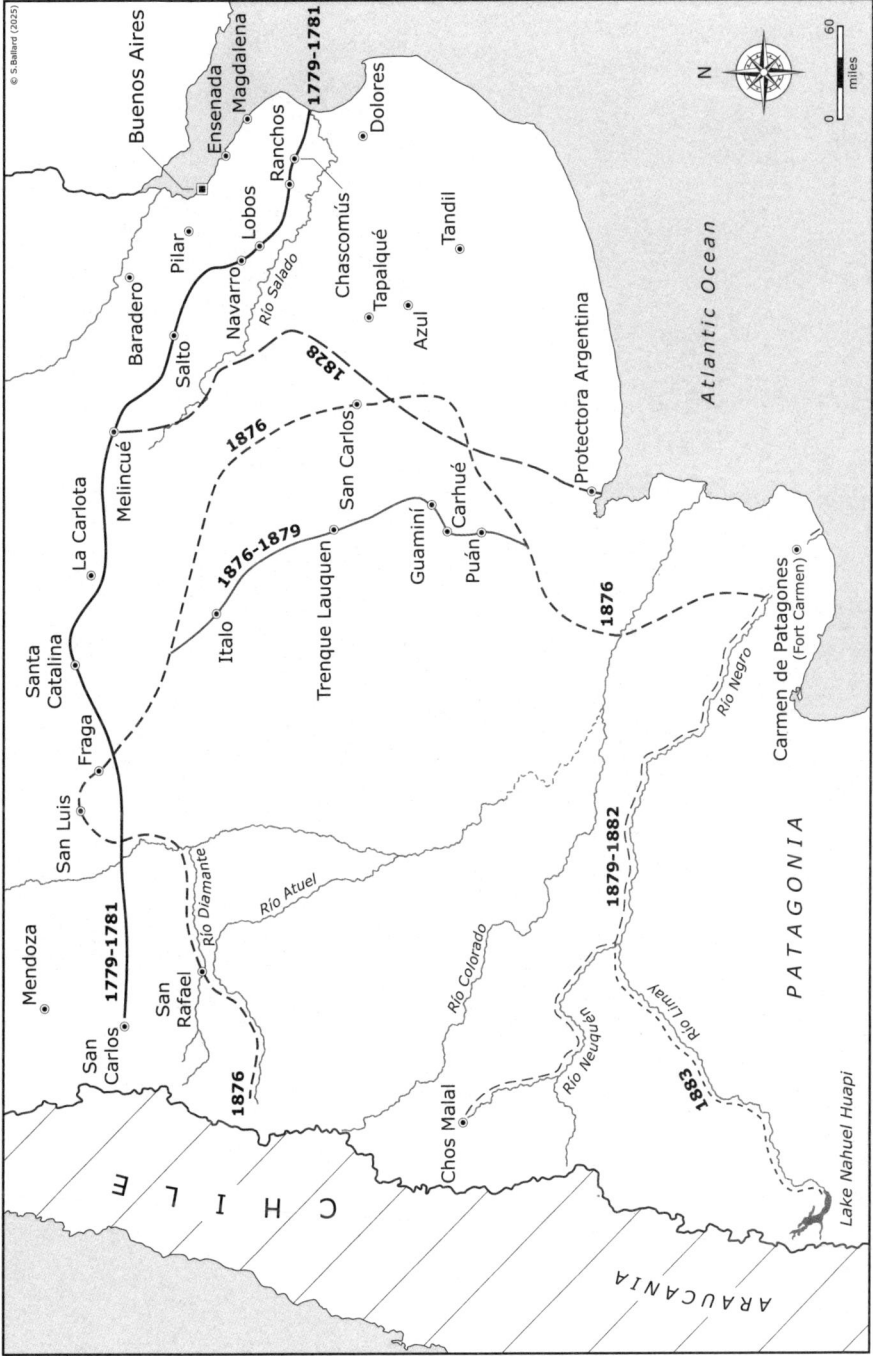

Map 3: 'The Conquest of the Desert': Argentina's territorial expansion

INTRODUCTION

AT THE WORLD'S END

Walk east out of Ushuaia, on the Isla Grande (Big Island) of Tierra del Fuego, and the coastal road soon gives way to a forest of birch and the southern beech trees known to scientists as Dombey's beech (*Nothofagus dombeyi*), which Argentinians call *coigue*. Some of these trees are hundreds of years old, and their creaking is the only sound you hear as you step round the rotting trunks that block the path or lean up against healthy trees like drunks in a bar. In this ancient forest, where everything has been growing and mouldering for centuries, it's the small details that catch your attention: the splashes of orange from the parasitical mistletoe known as *Farolito chino* (Chinese lantern); the twisted knots of fungi; the lime-green lichens that cling tenaciously to trees, bushes and rocks. Most of the time, you can barely see the sky through the canopy of leaves overhead, and then the path takes you on a cliffside overlooking the Beagle Channel, and looking back, you can see whole swathes of forest blown back by the fierce Patagonian winds, as if a giant comb had been dragged through it. On the opposite side of the channel, layers of green and russet leaves spread up the mountains to the permanent snowline on Navarino Island like a giant Christmas pudding, and further east, the mountains and the sea form a perfect V-shape at the entrance from the Atlantic.

On calm days, this legendary strip of water is as placid as an English river, and you might catch an occasional glimpse of a whale or sea lion as you stand on the edge of one continent looking out towards another. And then, suddenly, without any warning, all this can change, and you can find yourself running for shelter in a hailstorm, with the winds tearing at the trees and the waves crashing against the shore, and you get a sense of the 'universal

1

signs of violence' and 'savage Magnificence' that Charles Darwin observed in these forests when he saw them for the first time in 1832. On 17 December that year, the *Beagle* sailed through the Le Maire Strait between the Isla de los Estados and the southernmost tip of the Isla Grande, where Darwin observed 'scattered Indians' on the shore, who were lighting fires to announce the arrival of these unknown strangers to their neighbours on the other islands. On 18 December, Darwin went ashore with Captain Robert FitzRoy, the commander of the *Beagle*, and his officers to get a closer look. 'It was without exception the most curious & interesting spectacle I ever beheld,' he wrote in his journal that day:

> I would not have believed how entire the difference between savage & civilized man is. It is greater than between a wild & domesticated animal, in as much as in man there is greater power of improvement ... From their dress &c &c they resembled the representatives of Devils on the Stage, for instance in *Der Freischütz*.[1]

These 'devils' were members of the Haush people, a small tribe of hunter-gatherers who had inhabited the south-eastern tip of the Isla Grande for thousands of years, to whom Darwin, FitzRoy and the Beagle crew gave the generic name 'Fuegians'. Darwin encountered many 'savage' peoples in a five-year voyage around the world that took him from South America and the South Seas to New Zealand, Australia and South Africa. Of all these encounters, the 'savages' of Tierra del Fuego made the deepest impression. Summing up the highlights of his journey at the end of his journal, he contrasted the tropical forests of Brazil, 'where the powers of life are predominant', with 'those of Tierra del Fuego, where death & decay prevail'. Though both landscapes were 'temples filled with the varied productions of the God of Nature', Darwin was surprised to find that 'the plains of Patagonia most frequently cross before my eyes. Yet these plains are pronounced by all most wretched and useless. They are only characterized by negative possessions, without water, without trees, without mountains, they support merely a few dwarf plants.' What explained the enduring impact of these 'arid wastes'? Part of the answer lay in their inhabitants:

Of individual objects, perhaps no one is sure to create astonishment, than the first sight, in his native haunt, of a real barbarian, of man in his lowest and most savage state. One's mind hurries back over past centuries, & then asks could our progenitors be such as these? Men, whose very signs & expressions are less intelligible to us than those of the domesticated animals; who do not possess the instinct of those animals, nor yet appear to boast of human reason, or at least of arts consequent on that reason.[2]

Darwin continued to reflect on these differences throughout the five decades that followed his return to England in 1836. The 'savages' of Tierra del Fuego and Patagonia appear repeatedly in his research notebooks, in the books that defined his reputation, in letters to his family, friends and in debates with his scientific peers. And throughout the years in which Darwin pondered the questions he had first asked himself in these 'arid wastes', the territories he visited were brought into the orbit of modernity and 'civilization'. In 1870, English missionaries—inspired by his writings and those of FitzRoy—established an Anglican mission in Ushuaia. During a horseback ride into the Argentine interior in 1833, Darwin witnessed a 'war of extermination' between the army of Buenos Aires, under the command of General Juan Manuel de Rosas (1793–1852), against the Indigenous tribes of northern Patagonia and the vast grassy plains of central Argentina known as the Pampas. It was not until 1879 that the Argentinian army returned to these territories and completed the conquest that General Rosas had begun. By the time Darwin died in 1882, most of the Indigenous Peoples of mainland Patagonia had been killed or forcibly removed from the lands they occupied. Further south, the impact of 'civilization' on the 'savages' he encountered in Tierra del Fuego was no less devastating.

When Darwin first arrived in the Beagle Channel, as many as 10,000 people may have lived in the Isla Grande and the territories adjoining the Strait of Magellan. By the end of the century, the peoples whom Darwin and FitzRoy knew as 'Fuegians' had been decimated by violence and disease and acculturation into Chilean and Argentine society. This book is the story of the subjugation, dispossession and near-annihilation of the southernmost Indigenous Peoples of the Americas. At the same time, it is also a story of

3

'Darwin's savages'—of the people he saw, his responses to them and the conclusions he and others drew from these encounters in the scientific debates about race and racial difference, and human origins, which also extended to the countries these peoples had come from.

* * *

In both Chile and Argentina, the conquest of Patagonia is a well-known story. In the English-speaking world, this history is barely known outside the specialist departments of university campuses. Thousands of pages have been written about the destruction of Tasmania's Aboriginal population, about the genocide of the Herero and Nama peoples in South West Africa or King Leopold's Congo Free State, but the subjugation of the Mapuche, Tehuelche and other Patagonian First Peoples is little known. After so many books, films and television series, Sitting Bull, Crazy Horse and Geronimo are household names in many countries around the world, but the long struggle between white settlers and the Indigenous caciques (chiefs) of Patagonia and the Pampas has received little attention outside Argentina. Custer's 'last stand' of the Seventh Cavalry in 1876 is one of the most famous events in American history, but the scale of that battle is dwarfed by the huge raids carried out by Indigenous tribal coalitions in Buenos Aires Province in the 1870s.

In 1876, the Argentine government built a 250-mile-long fortified ditch in an attempt to prevent these raids—the only time in history that any modern state has undertaken such measures against Indigenous opponents. Very little of this has filtered out into the wider world, even though the conquest of Patagonia was an international as well as a national enterprise. Between 1850 and 1900, English, Welsh, Spanish, Scottish, Croatian and German immigrants all took part in the colonization of Patagonia and Tierra del Fuego, in an era in which Europeans and the descendants of Europeans gained control of some 40 per cent of the earth's surface. European, particularly British, capital played a decisive role in bringing Patagonia into the international economy. This period of history was also the apotheosis of 'scientific racism', in which European scientists concerned with racial differences and the 'varieties of man' saw the 'savage' and 'primitive' peoples of the world as historical anachronisms, whose backwardness condemned them to historical obsolescence.

In his seminal anticolonial polemic *Discourse on Colonialism* (1950), the Martinican poet Aimé Césaire rejected the association between 'colonization and civilization' through which European colonial powers justified their domination of the 'barbaric' peoples of the world. Césaire reserved special contempt for the 'psychologists, sociologists *et al.*' for their 'rigged investigations, their tendentious speculations, their insistence on the marginal "separate"' character of non-white peoples, which had enabled them 'to impugn on higher authority the weakness of primitive thought'.[3] Patagonia's First Peoples also became the object of these 'investigations' in the nineteenth century, both in Europe and in Chile and Argentina, and much of this interest can be traced to the most influential scientist of his times, who did more than any single writer to bring Patagonia to the attention of the international scientific community.

* * *

Today, any visitor will very quickly become aware of Patagonia's associations with Charles Darwin. Mountains, fjords and rivers retain the names given to them during the *Beagle*'s three voyages. Towns, hotels, ski lodges, trees and fungi have been named after Darwin or FitzRoy. Travel agencies offer 'Darwin experience' kayaking trips along the Río Santa Cruz that Darwin and FitzRoy once explored, and cruise ships follow 'Darwin's route' through the Beagle Channel and the Strait of Magellan. Images of the youthful naturalist, and the venerable bearded sage he later became, can be found in hotels, shops, restaurants and ferries. To some extent, Darwin's ubiquitous presence is an exercise in touristic branding, in its attempt to define Patagonia as a place of intellectual and physical adventure. But the Darwinization of Patagonia is also a testament to Darwin's fundamental role in 'unveiling' Patagonia in the nineteenth century.

Travellers, missionaries, explorers, anthropologists, naturalists and soldiers followed in Darwin's footsteps. In a 1946 book on Rosas's campaign in the south, an editorial note described Darwin's journeys and writings as the decisive element in establishing Argentina's authority over the Indigenous territories of the south. Its author reminded his readers that '[t]he work of Darwin was unequalled by the contemporary men of science. It was intensely read in the two worlds, and demonstrated with its testimony that the Patagonian lands were subject to the authority of the Argentinas.'[4] It is questionable whether Darwin consciously intended any such thing. Nevertheless, this association between military conquest and science—particularly Darwinian science—was a recurring theme in the colonization of Patagonia. Nineteenth-century Argentinian scientists discovered their southern territories through Darwin's and FitzRoy's journals. Ethnologists, anthropologists and museum curators followed the Argentinian army southwards and ransacked Indigenous graveyards in search of skulls and skeletal remains for European and Argentinian museums. At the La Plata Natural History Museum, Indigenous prisoners from the 'Conquest of the Desert' were kept in the basement as living exhibits and repositories of ethnological knowledge, some of whom went on to become permanent exhibits in museum showcases after their deaths. Living 'Fuegians' were brought to European 'human zoos', where they were measured and studied by some of the leading scientists of the day. Through conquest, Patagonia became a field of science. Armchair scholars who had never been to Patagonia formed hypotheses about the world's 'savage' and 'primitive' peoples based on Darwin's depictions of Tierra del Fuego's inhabitants in the discussions of the 'varieties of Man' that dominated scientific debate in the second half of the nineteenth century.

Darwin also participated in these debates, and he often cited his experiences of Patagonian 'savages' in his contributions to the discussions about racial difference and human genealogy that dominated Victorian science. Darwin's attitudes to race and the impact of his own ideas on nineteenth-century racial science have long been subjects of scholarly and not-so-scholarly discussion. Some writers, often but not exclusively with a creationist orientation, have attempted to trace a direct link between Darwinism and Nazi racial science.[5] Historian of science John Desmond Bernal argued that Darwinism—regardless of Darwin's own intentions—replaced religious explanations for 'the dominance of classes or races' with rational, scientific and 'biological' explanations for historical processes that justified 'the ruthless exploitation of man by man, the conquest of inferior by superior peoples'.[6]

Darwin's many defenders point out that his ideas were appropriated for uses he had not intended or cite his fervent opposition to slavery to repudiate accusations of racism. Two of Darwin's biographers have even suggested that his science was driven by anti-racism.[7] In her study of Victorian anthropological writings, literary critic Gillian Beer warns against the tendency 'to dismiss all Victorian writers as racist because they use vocabulary that offends us now, or because they all work within a developmental view of history', not least because such dismissal 'has the effect of absolving present-day readers and allowing us to feel enlightened. The rejection costs us no inquiry.'[8]

No one could reasonably dispute this, but it should be possible to scrutinize such writers without feeling complacent about our own moral progress, and there is no reason why Darwin's attitudes to race should be exempt from such enquiry. The late palaeontologist and historian of science Stephen Jay Gould once argued that Darwin's 'paternalist' contempt for Tierra del Fuego's 'savages' was cancelled out by his fervent abolitionism.[9] There is no doubt that Darwin's visceral responses to the 'Fuegians' are very different from his more positive assessments of Brazilian slaves, whom he frequently described in sympathetic and humane terms that followed logically from his abolitionist views. But in the nineteenth century, it was perfectly possible to oppose slavery while also regarding the world's Indigenous Peoples as inferior or retarded versions of humanity, whose inferiority condemned them to 'extinction' through their interactions with a superior white civilization.

Darwin frequently expressed such views, and he often cited his own experiences of Patagonian 'savages' to explain or justify them. A kind and humane man, who was appalled by the 'war of extermination' he witnessed in the Pampas, he repeatedly used the word 'extermination' in his discussions of 'inferior' races with a troubling ambiguity and lack of clarity. We do not have to become what the Argentinian philosopher Héctor Palma calls 'implacable inquisitors regarding Darwin's passive acceptance of generalized and extended beliefs' in order to explore these assumptions and prejudices.[10] If Darwin was a man of his times, then we can also see what he took from his times and what his times took from him, and we can also consider the extent to which science has changed in its treatment of race and racial difference.

* * *

The consequences of this history are still being played out in Argentina and Chile. In his popular travel book *The Old Patagonian Express* (1979), Paul Theroux described his relief on arriving in Argentina to find himself in a country where the Indians he had seen in Peru and Bolivia were absent. In Argentina, the faces he saw from his train were 'the faces one might see on any train in the United States, or Europe for that matter. It was possible to enter a crowd in Argentina and vanish.'[11] Buenos Aires was similarly 'a city without a significant Indian population', apart from Indians from Paraguay or Uruguay, 'who worked as domestics, ... lived in outlying slums, [and] were given little encouragement to stay'.[12] The Buenos Aires he saw was the one that Argentinians have tended to see—the sophisticated capital of a 'white Argentina' made up of European immigrants who 'came from the boats' from the late nineteenth century onwards. During the celebrations that followed Argentina's victory in the 2024 Copa América, the Argentinian football team caused a minor diplomatic incident by singing a racist song mocking the African origins of certain members of the French national team. The damage was compounded when Argentina's right-wing vice-president, Victoria Villarruel, posted in support of the team that Argentina, unlike France, had never had 'colonies or second-class citizens'. Villarruel also described Argentina as a country built on the 'sweat and courage of the Indians, the Europeans, the *criollos*, and the negroes'.[13]

To call this ill-judged post a distortion of history would be to understate it considerably. Patagonia's conquest was as much a process of 'internal' colonization as the conquest of the American West; its protagonists did not regard the 'sweat and courage of the Indians' as a hallmark of Argentinian identity. On the contrary, the eradication of Indigenous Argentina was seen as an essential precursor to the creation of the white Europeanized nation that

Theroux admired. Today, we no longer take the distinctions between 'savage' and 'civilized' peoples for granted, and Argentinians no longer see the conquest of Patagonia as a victory for civilization over the barbaric 'desert' of the south. In Argentina, and Chile, nineteenth-century colonial history has become the object of the same statue wars and 'battles for memory' that have unfolded across the world in the twenty-first century. 'Why do we never speak of Indians in Argentina?' asked the Argentinian author David Viñas in his classic study *Indios, ejército y frontera* (Indians, army and frontier, 1982). 'What does it mean that Indians have been pushed offstage, relegated to ethnology and folklore or, even more sadly, to tourism and local colour pieces in the press?'[14]

Today, Argentinians do speak of Indians, and Indians also speak for themselves, in both Argentina and Chile. In both countries, First Peoples have re-emerged as protagonists in campaigns for land and civil rights and *plurinacionalidad* (plurinationality). In writing about Darwin's 'savages', I have often felt the twenty-first century looking over my shoulder, in a new century in which the peoples who were once described as savages and barbarians are seeking justice, equality, historical reparation and restitution, and these struggles will also feature in the pages that follow.

PART I

DISCOVERY

Perhaps there is no other region in the world about which so much has been spoken and so little is known than Patagonia.

Alcide d'Orbigny, *Voyage pittoresque dans les deux Amériques* (Picturesque voyage to the two Americas, 1836)

1

RES NULLIUS

For centuries, Patagonia's reputation as a mysterious, inhospitable wilderness at the world's southernmost limits derived almost entirely from what was written and said about it by people who knew little or nothing about it. In 1822, the American explorer and sealer Benjamin Morrell wrote of the 'many contradictory statements and monstrous absurdities—fables more difficult to believe than Gulliver's Travels' that had ensured that 'the character, history, and everything relating to Patagonia are still involved in great obscurity'.[1] These fables begin on 31 March 1520, when the Portuguese navigator Ferdinand Magellan wintered at present-day Port San Julián in Argentina during his 1519–22 circumnavigational voyage. According to the colourful account of Magellan's Italian chronicler Antonio Pigafetta, Magellan's crew had spent two months in the bay when they saw 'a naked man of giant stature on the short of the port, dancing, singing, and throwing dust on his head'. Magellan invited the 'giant' to meet with him, only to find, according to Pigafetta, that '[h]e was so tall that we reached only to his waist ... his face was large and painted red all over, while about his eyes he was painted yellow'.[2]

Magellan offered the 'giant' some food and showed his guest a 'large steel mirror', which terrified him so much that he 'jumped back, throwing three or four of our men to the ground'. Magellan was visited by other 'giants' over the next few days, one of whom he baptized and named John. Two other 'giants' were tricked into wearing leg irons and taken prisoner. One escaped, and the other was wounded. When Magellan's men tried to capture the wives of these 'giants', their companions tried to free them, and one of

13

Magellan's sailors was killed with an arrow. Pigafetta called their country 'Patagonia', and his map of Magellan's voyages refers to the 'Regione Patagonica' divided by the 'Streto patagonico' (Patagonian Strait). For centuries, it was believed that the word Patagonia derived from the Spanish word 'pata' (foot), in reference to the big feet of these 'giants'.

In 1952, the Argentine philologist María Rosa Lida de Malkiel argued convincingly that Magellan had taken the name from the fifteenth-century Spanish chivalric romance *El Primaleón*, whose protagonist battles the doglike monster Patagón.[3] In Lida's view, Magellan gave this name to the Indians he encountered not because of their size but because of their perceived wildness and ugliness. As late as 1981, however, the Chicago edition of the *Encyclopaedia Britannica* still claimed that '[t]hese people, from whom the name of Tierra de Patagones was given by Magellan on observing their large footprints, are remarkable for their great stature, having an average height of 6 ft. to 6 ft. 4 in.'

* * *

Other Patagonian myths and fables were equally enduring. In the sixteenth century, Spanish conquistadores made occasional expeditions into the Patagonian interior in search of the 'City of the Caesars', also known as the 'Wandering City' or 'Trapalanda', that was reputedly paved with silver and gold and inhabited by a race of blue-eyed, blonde-haired people who wore three-cornered hats. The English naval officer James Weddell (1787–1834) hypothesized that the language spoken by the Indigenous Peoples of Tierra del Fuego contained Hebrew words derived from Spanish, which had brought 'Arabic, the language of the Moors' to Tierra del Fuego.[4] In Jules Verne's novel *Les enfants du capitaine Grant* (The children of Captain Grant, 1867–8), a map found in a shark's belly inspires an epic rescue expedition in search of the lost sea captain of the shipwrecked commander of the *Britannia*, Captain Grant, that leads its protagonists into Patagonia in the belief he has been captured by Indians. In a 1922 British comic strip, the detective Sexton Blake travels to Patagonia in search of Patagonia's 'Loch Ness monster'—a fictional variation of a real-life Argentinian expedition that took place in search of a cryptid monster known as the Patagonia Plesiosaur that inhabited a lake in northern Patagonia. The British author Bruce Chatwin was inspired to write his seminal travel book *In Patagonia* (1977) by a 'piece of brontosaurus' hanging in his grandmother's dining room.

Patagonia's size, remoteness and inscrutability, and its location at the 'end of the world', easily lent itself to stories of giants, monsters and imaginary cities. For hundreds of years, the only European visitors to Patagonia were whalers, sealers and seamen on their way from one ocean to the other, few of whom ventured inland from the coast. Port Famine, Desolation Land, the Brunswick Peninsula, Port Sir Francis Drake, Elizabeth Bay, Port Desire, Louis Island, Island of the Apostle—these are the names through which Patagonia first appears in the European imagination. Beyond these coastal waymarks lay a *res nullius*—'property of no one'— about which little or nothing was known. In the 1562 map drawn up by the Spanish cosmographer Diego Guttiérez, the 'Tierra de Patagones'—'Land of the Patagons'—appears as a 'Gigantum Regio' or 'Region of Giants' that encompasses all the territories south of the River Plate, including present-day Chile and Argentina but without the Andes in between. A 1746 Spanish colonial map of Patagonia is almost entirely blank, save for a few topographical features and a warning that most of the territory is 'inhabited by Infidels' who 'live wandering, without towns, without a fixed place, without cemeteries'. Other maps refer to 'Indian territory', 'uncontrolled territory' or 'sterile lands'. As late as 1838, a map by the London-based Society for the Diffusion of Useful Knowledge depicted the Patagonian interior as a large blank space, 'Inhabited by Wandering Tribes of Indians.'

The arduous and dangerous voyage around Cape Horn or through the Strait of Magellan did little to encourage Europeans to look any further. In 1584, the Spanish explorer Pedro Sarmiento de Gamboa left 400 men and thirty women with eight months' provisions at Port Famine, on the north bank of the Strait of Magellan, with the intention of founding a permanent Spanish settlement. Two years later, the English mariner Thomas Cavendish arrived at the same spot to find that all but twenty-three survivors had died of starvation 'like dogges in their houses, and in their clothes' in a ghost town 'wonderfully taynted with the smell and the savour of the dead people'.[5]

The Dutch philosopher and geographer Cornelius de Pauw devotes an entire chapter of his *Recherches philosophiques sur les Américains* (Philosophical researches on the Americans, 1770) to Patagonia, which he called 'the most inhospitable and unfruitful of all the regions of the earth', even though he had never set foot in the Americas.[6] 'The country about this place has the most dreary and forlorn appearance that can be imagined,' wrote the British naval officer Captain Samuel Wallis in 1766 of a voyage through the Strait of Magellan that recalled 'the ruins of Nature devoted to everlasting sterility and devastation'.[7] More than half a century later, Captain Pringle Stokes, FitzRoy's predecessor as captain of the *Beagle*, described the Tierra del Fuego archipelago as cheerless:

> Nothing could be more dreary than the scene around us. The lofty, bleak, and barren heights that surround the inhospitable shores of this inlet were covered, even low down their sides, with dense clouds, upon which the fierce squalls that assailed us beat, without causing any change. They seemed as immovable as the mountains where they rested ... as if to complete the dreariness and utter desolation of the scene, even the birds seemed to shun its neighborhood.[8]

* * *

Variations on this theme of desolation recur repeatedly in descriptions of the Patagonian mainland and Tierra del Fuego. In 1979, the great writer and poet Jorge Luis Borges told Theroux that Patagonia was a 'dreary place, a very dreary place'—a description Theroux found confirmed in the 'nothingness' of the countryside around Esquel, with its 'prehistoric look, the sort that forms a painted backdrop for a dinosaur skeleton in a museum, this place had no landmarks, or rather it was all landmarks, one indistinguishable from the other— thousands of hills and dry riverbeds'.[9] In 2001, the late Chilean novelist Roberto Bolaño jokingly compared Patagonia to 'the vast and shifting territory of the Mexican–North American frontier. Instead of desert, pampa; instead of villages sleeping in the sun, shacks beaten by the sun and the southern rains.'[10]

This picture of monotonous windswept desolation does not do justice to the variety of Patagonian landscapes. Drive south down Argentina's legendary Route 40, which runs parallel to the Andes, and you will soon find the arid steppes covered with wild yellow grass and the shrubland bushes known as *mata negra* that fully bear out the descriptions of Patagonia as a barren desert. But then these plains give way to enormous glacial valleys ringed by flat-topped mountains that never seem to come any closer; to alpine forests that would not be out of place in Switzerland or Bavaria; to the bronze foothills and white peaks of the Andean cordillera; to glaciers, icefields and turquoise lakes that look like inland seas. To a traveller coming from crowded, domesticated Europe, the vastness of Patagonia suggests an ancient pre-human world, and this sense of space is reinforced by its immense skies lined with extended cirrus cloud formations, by its fiery sunsets that extend for miles and more than compensate for even the most monotonous terrain. Further south in Chile, the windblown plains between Puerto Natales and Punta Arenas merge into the labyrinth of channels, fjords and tree-covered islands that make up the southern tip of Chile's 'crazy geography', where the land looks as if it were broken up by a giant hammer.

Today, even Patagonia's more remote areas have been mapped, named and divided into administrative districts. Argentine Patagonia consists of more than 550,000 square miles, located between latitude 39° and 55° South, divided into five provinces, flanked by the Andes to the west and the Colorado and Negro rivers to the north, which run down from the Andes and across the country to the Atlantic about 45 miles south of the coastal city of Bahía Blanca. Nearly 2,000 miles separate Buenos Aires by car from the southernmost Argentinian city of Ushuaia—further than the journey from London to Moscow—across 260,00 miles of steppe and desert. Add Chilean Patagonia, and the combined territories amount to a million square kilometres. Chilean Patagonia begins at Puerto Montt at latitude 41° South, though some experts place the boundary at Temuco, some 220 miles further north in 'Norpatagonia', roughly in line with the Argentinian city of Neuquén.

Most of Chilean Patagonia is located in the southernmost provinces of Aysén and Magallanes, in addition to the western half of the Isla Grande and the islands of Tierra del Fuego across the Beagle Channel to the Isla Navarino in Chilean Antarctica. Today, the combined population of Chilean and Argentinian Patagonia is about 2 million, most of whom live in a few large towns and cities. When Darwin, Alcide d'Orbigny and the other European explorers visited Patagonia at the beginning of the nineteenth century, it would have looked much the same as it had done thousands of years before their arrival. But even if Patagonia was the 'property of no one' according to European legal standards, it was not empty.

* * *

The Cueva de las Manos—'Cave of Hands'—is about an hour's drive from the town of Perito Moreno, in the Patagonian province of Santa Cruz. An unpaved road takes you through bare rocky hills and yellowed scrubland, past small flocks of guanaco and the occasional ostrich-like bird that Argentinians call ñandu (Darwin's rhea) to the canyon overlooking the Río Pinturas. From there, you walk down to the fertile riverbed where the Río Pinturas sluggishly wends its way through a line of willow trees, flanked by ignimbrite cliffs, rocks and overhangs of gleaming bronze that would not look out of place in a John Ford movie. As you descend, you half expect to see a Toxodon or giant armadillo sipping at the river. It's not until you climb up to the walkway on the opposite side of the canyon that you can see the clusters of red, yellow and white hands, superimposed over each other, that hunter-gatherers stencilled on to the rock through stone blowpipes some 13,000 to 9,500 years ago.

No one knows why Palaeo-Indians made these collages as they followed the grazing animal herds along the Pinturas River, but archaeologists believe that the more recent hands may have been made by descendants of the Tehuelche people—the dominant Indigenous inhabitants of the Patagonian plains for thousands of years. The word 'Tehuelche'—a Mapuche Indian word meaning 'wild' or 'fierce' people—is the collective term historians and archaeologists use to describe the mosaic of tribes, clans, lineages, ethnic divisions and sub-divisions that once extended from the Gununa-Kena or northern Tehuelche who lived around the Colorado River and the Río Negro to the Aónikenk or southern Tehuelche who inhabited the plains below the Bahía San Julián and immediately above the Strait of Magellan.[11] Unlike the Incas or the Aztecs, the Tehuelche left no cities, roads or monuments. They were nomadic hunter-gatherers who depended for their food, clothing and shelter on the guanaco herds that once abounded in Patagonia. In addition to meat, the guanaco provided blankets and the characteristic Tehuelche cloaks known as *kais* or *quillangos* made up of two skins sewn together by sinews from the ostrich-like flightless rhea bird known as the *ñandu* and coverings for the *toldos*—tents—that the Tehuelche assembled and disassembled as they followed the seasonal movements of the herds.

The Tehuelche also hunted deer, foxes and pumas, using bows and arrows, spears and *boleadoras* (bolas)—grooved stone balls attached to ropes and cords, which were spun and thrown like a lasso to hamstring their prey. By the eighteenth century, the Tehuelche had begun to make use of the herds of wild horses introduced by the Spanish that spread across the Pampas following the foundation of Buenos Aires in 1536. Horses revolutionized Tehuelche society. They made it possible to transport their camps across greater distances and provided meat and blood to drink in the absence of water, as well as bones, hides and tendons to make household utensils and colt skin boots. Horses also changed how the Tehuelche made war. They soon became expert riders and dispensed with bows and arrows in favour of the bola, the knife and the 18-foot-long cane lances that they used with great dexterity.

Tehuelche *tolderías*—encampments—usually consisted of some thirty to forty families, living in tents that housed about ten people under the leadership of an elected chieftain or *cacique*. Their marriages were polygamous: men were allowed to have as many wives as they could support, and women could leave their husbands for a more powerful man. While the men hunted and made war, women assembled and disassembled the tents, took care of cooking, gathering wood and other household tasks and skinned and scraped the hides of the animals they caught, sewing them together into cloaks and blankets, which they decorated with elaborate interlocking geometrical motifs that demonstrated the caste and status of their owners.

European travellers describe the Tehuelche as long-limbed, gaunt and generally tall people, even if they were not the 'giants' described by Pigafetta. The men plucked their beards but rarely cut their long, plaited hair. They were fond of cards and dice. They feared the evil spirit called the Gualichu and the giant of the mountains called Trauco. They believed in a supreme being—the creator god Kóoch, who lived alone on the edge of the world and created the sea, the sun, the moon and the stars to relieve his solitude. They paid homage to the man-god Elal, who flew to Patagonia from a 'Big Island' on a swan's back, bringing with him the guanaco, the puma and other animals that would become his 'faithful friends'. The Tehuelche relied on shamans or *curanderos* (medicine men) assisted by elderly women with spiritual authority known as *machis* to treat the sick. These *machis* painted their faces with flour and starch and wore silver bracelets round their ankles as an indication of their status, and they were considered to have the ability to invoke and banish spirits, regulate sexual behaviour and treat illnesses with herbal remedies.

Some of these cures were challenging, by modern standards. In some cases, the *machi* relieved pain by opening a patient's ribs and cutting off a piece of their liver, which was then given to them to eat. Other treatments involved the ritual slaughter of a sheep and a colt in a shady grove, after which the patient was taken to lie with the animals while the medicine man and the *machi* danced around them. Afterwards, the medicine man sucked blood from the painful area and spat it out in the direction of the sun, before the patient was anointed with the blood of the animals and the dances resumed, sometimes with the patient's participation. The Tehuelche believed in the afterlife and buried their dead with their weapons, horses and dogs—and sometimes their wives—to accompany them to the next world, before burning all their earthly belongings except their tents.

Authority in the Tehuelche community rested in the hands of the chief, or their subordinate *capitanejos* (lieutenants). The cacique decided when to break camp and the direction of travel and presided over matters of war and peace. They were usually men, though women could also become caciques, such as the Tehuelche chief María who helped FitzRoy and his crew during the *Beagle*'s first voyage. These camps were sometimes organized into larger groupings, in which different tribes and lineages formed semi-permanent *cacicatos* or chieftainships that held sway over a particular territory. Their members might also attend councils with caciques from other Indigenous groups at councils or *parlamentos* (assemblies), in which hundreds and sometimes thousands of Indians gathered to organize military alliances, elect a commander-in-chief and resolve matters of war and peace or trading relationships between different peoples.

Scattered across such a vast territory and often isolated for long periods, Tehuelche bands inevitably interacted with other First Peoples. In north-west Patagonia, they traded with the forest-dwelling Pehuenche or 'pine eaters', for whom pine nuts were a staple of their diet, and also with the Picunche or 'north people' who lived in central Chile and the present-day Neuquén Province in northern Patagonia. Tehuelche bands lived alongside the Puelche, or 'people of the east', who lived on the lower slopes of the Andes and in the triangle formed by the Neuquén and Limay rivers. In 1670, the Jesuit priest Nicolás Mascardi travelled to Lake Nahuel Huapi, near present-day Bariloche, where he identified at least two separate groups of Puelches living around the lake who spoke two different languages, in addition to an entirely different tribe called the Poyas, who spoke their own language. Some of these Puelche groups were primarily hunters. Others were expert fishermen and traders, who were able to disassemble and reassemble their bark canoes, which they carried across the Andes to the island of Chiloé on the Pacific Ocean, where they traded with the Huilliche people.

From the late sixteenth century onwards, the native peoples of north-west Patagonia and the Pampas came under the influence of the Mapuche, Huiliche and Picunche peoples, to whom the Spanish gave the collective name 'Araucanians' or 'Aucas'. Unlike the Indigenous Peoples east of the Andes, the Mapuche of central Chile lived in villages and fixed settlements, where they combined the cultivation of maize, sweet potatoes, melons and other fruit and vegetables with livestock raising, llama wool production and the manufacture of blankets and ponchos. For centuries, the Mapuche had resisted Spanish colonization with more success than any other Indigenous society in Latin America, to the point that Araucanía

21

was effectively an autonomous territory from the early seventeenth century to the late nineteenth, bounded by the Biobío and Toltén rivers.

Between the sixteenth and early nineteenth century, Mapuche, Huilliche and Pehuenche bands and clans crossed the Andes to raid Spanish settlements in the Pampas and northern Patagonia, in the process historians call the 'Araucanization of Patagonia'. In some cases, the Mapuche displaced or subjugated the Indigenous Peoples who already inhabited these territories. Other groups were drawn into the Mapuche cultural orbit and adopted Mapudungun—the language of the Mapuche—as a common language. These 'Araucanized' Indigenous groups included the Ranquel or Rankülche—'people of the reeds', made up of Mapuches, Pehuenches and creole military deserters, who settled in the southern Pampas and northern Patagonia in the late 1700s. Spanish colonial maps often refer to such groups as 'wandering Indians', but they tended to follow familiar trading and hunting routes and *rastrilladas* (rake-trails) that marked the most common routes from southern to northern Patagonia or the 'Chilean road' that connected the Atlantic coast to the Pacific along the line of the Río Negro. Nor were they entirely nomadic. Some *tolderías* were permanent or semi-permanent. In 1782, the explorer Basilio Villarino (1741–85), a pilot in the Spanish Royal Navy, navigated the Negro, Limay and Collón Curá rivers in northern Patagonia and reached the territory between Lake Nahuel Huapi and Junín de los Andes, which the local Indians called the 'Country of Wild Apples' because of its profusion of apple orchards. 'Where fruits are picked, you can always find something in the trees that the pickers have overlooked,' Villarino observed, 'but the Indians are such careful pickers, that nothing goes unnoticed.'[12]

* * *

The First Peoples of the far Patagonian south developed very differently. Separated from the mainland by the Strait of Magellan, the inhabitants of Tierra del Fuego were unaffected by the cultural shifts on the mainland. The Haush or Mane'kenk people whom Darwin encountered on the south-eastern tip of the Isla Grande probably never numbered more than 1,000. They lived by foraging, hunting and fishing, protected from the elements in the crude 'wigwams' built from branches, foliage and otter or seal skins that Darwin observed in the first half of the nineteenth century. Further north, the hills, forests and plains north of contemporary Ushuaia were the domain of the Selk'nam or Ona 'foot-Indians', who hunted guanaco and other animals with bows and arrows.

Further south, the coasts of Tierra del Fuego, Navarino Island and the Wollaston Islands north of Cape Horn were inhabited by the nomadic Yamana or Yagán people—'canoe Indians' who navigated the fjords and canals in small bark canoes.[13] The islands between the Brunswick Peninsula and the Strait of Magellan were home to the Kawésqar or Alacalufe, another seafaring people who moved between the mainland and the islands of the south-west. All these peoples spoke their own languages and had their own rituals, myths and belief systems. Though the Selk'nam wore their characteristic guanaco cloaks, the Kawésqar and the Yagán were often naked or semi-naked, apart from a loin cloth and a small otter-skin over their shoulders and a layer of seal or whale oil to keep warm. Both peoples were expert navigators, and their canoes were often paddled by women. Few men in Tierra del Fuego could swim, but girls were dipped in cold and sometimes freezing water even as toddlers so they could follow the example of their mothers, diving for shellfish and tending the fires that were kept alight at all times and carried even in their canoes on a bed of sand. Fires were used for warmth as well as for sending messages and signals—a custom that led Europeans to dub their territory Tierra del Fuego: the Land of Fire.

These are only some of the 'various nations of Indians, barbarous and ignorant of Spaniards' identified on a 1786 map of Patagonia. Others disappeared long before any maps of the region were ever drawn up. In written terms, Patagonia's First Peoples do not enter history until the Spanish colonial empire first began to turn its attention to the 'undiscovered' territories at the far end of the American continent.

2

PATAGONIAN ENCOUNTERS

Pigafetta's half-comic, half-tragic description of Magellan's meeting with Patagonian 'giants' contains many of the elements that were repeated in subsequent encounters between Europeans and Patagonian First Peoples: initial incomprehension, astonishment and curiosity, followed by mimicry, attempts at friendship and the ever-present subject of food. In 1619, two Spanish caravels anchored at a bay they called Bahía Buen Suceso—Bay of Good Success—in the Le Maire Strait, where they were welcomed by a group of Haush people, who 'assisted them in wooding and watering, without betraying the least distrust'. In 1778, the French admiral, explorer and veteran of the Seven Years War Louis-Antoine de Bougainville (1729–1811) invited a group of Fuegian 'savages' on to his ship, where '[w]e made them sing, and dance, let them hear music; and, above all, gave them to eat, which they did with much appetite'.[1] In 1822, Weddell extended a similar invitation to a group of Fuegians, where 'one merry fellow of our crew commenced singing and dancing, at which the Fuegians formed a circle round him, and imitated his song and dance most minutely'.[2]

These amicable interactions could turn sour very quickly. Weddell had one Fuegian flogged for stealing. In 1578, Francis Drake and his crew landed at Bahía San Julián, where they met about thirty 'country people', whom they plied with gifts of beads and coloured glasses and entertained with dances and music. A few days later, a bloody skirmish took place when the 'Patagons' attempted to prevent another boat from coming ashore. In June 1578, Drake and his crew engaged in 'cruel and bloody exchanges' with a group of Indians in the Strait of Magellan.[3] In December 1586, the

English Mariner Thomas Cavendish and his crew anchored in a bay they named Port Desire, where they were shot at by Indians who 'seldome or never see any Christians: they are as wilde as ever was a bucke or any other wilde beast'.[4] In 1599, a flotilla of Dutch ships spent five months stranded in the Strait of Magellan due to adverse winds, where more than 120 crewmembers died from hunger, exposure or attacks by Indians. One of these ships was commanded by Captain Sebald de Weert, who described how his half-starved crew opened fire on seven canoes filled with naked 'giants' and suffered various casualties before the Indians retreated. At Penguin Island, De Weert's crew found a 'Patagonian woman among the rocks' whose 'numerous tribe' had been massacred by the Dutch filibuster Olivier van Noort in another voyage that same year.

In February 1624, seventeen Dutch sailors from the Nassau fleet commanded by Admiral Jacques L'Hermite were beaten to death by a party of Yagáns during an overnight stay on Navarino Island, in what may have been revenge for Van Noort's massacres in 1599. In a later account of this incident in the Scottish plagiarist John Callander's *Terra Australis Cognita* (Voyages to the Terra Australis, 1766), one of the survivors of the Nassau fleet described the perpetrators as 'rather beasts than men; for they tear human bodies to pieces, and eat the flesh, raw and bloody as it is'.[5] This episode was the first association between the Fuegians and anthropophagy, and like Patagonia's 'giants', it became one of the defining images of the barbaric and irredeemable savages lurking on the fringes of what had become a vital maritime thoroughfare in Europe's trade routes between the Atlantic and the Pacific oceans.

* * *

These intermittent confrontations were mostly confined to the coast. Europeans had little interest in penetrating any further inland. Until the late eighteenth century, Patagonia was nominally part of the Captaincy-General of Chile to the west and the Río de la Plata territories east of the Andes that belonged to the sprawling Viceroyalty of Peru. Unable to conquer the Araucanian territories of central Chile, the Captaincy-General had neither the ability nor the inclination to venture any further south. Spanish interest in the

Río de la Plata region was largely due to the role of the port of Buenos Aires as an export hub for cereals, cattle hides and dried beef to the slave plantations of Brazil and the Caribbean.

In the eighteenth century, cattle ranchers from Buenos Aires began to push southwards, in search of the herds of wild 'cimarron' cattle that proliferated in the Pampas. By the late seventeenth century, these herds had become so abundant that the 'cowboys' known as gauchos would slaughter a cow just to eat a single meal, leaving the rest of the carcass for dogs and wild animals. Until the eighteenth century, the *estancieros*—ranchers—of Buenos Aires were able to find cattle without venturing too far from Buenos Aires and bring their animals to be slaughtered in the city's giant slaughter-yards that Darwin later visited. By 1715, these herds were becoming scarcer, and the more daring ranchers began to encroach on the territories that 'wandering Indians' regarded as their ancestral grazing lands. In *A Description of Patagonia* (1774), the English missionary, physician and naturalist Thomas Falkner (1707–84) describes a cacique named Cacapol who led 4,000 Tehuelches, Huilliches and Pehuenches in a series of raids on white settlements south of Buenos Aires in 1739 'with so much judgment, that he scoured and dispeopled, in one day and a night, about twelve leagues of the most populous and plentiful country in these parts'.[6]

When a Spanish militia massacred the inhabitants of an Indian camp in response to a raid on the settlement of Luján, 300 Indians attacked the settlement a second time and 'killed a great number of Spaniards, took some captives, and drove away some thousands of cattle'. An even larger militia-army attacked a group of peaceful Huilliches and a Tehuelche *tolderia* that supposedly enjoyed the protection of the governor of Buenos Aires. In response, Indians attacked the Spaniards, 'at once, down the whole length of the River of Plate, on a frontier of a hundred leagues', until the government of Buenos Aires sued for peace. By the late eighteenth century, this pattern of raids, massacres, counter-raids and treaties was well established in the southern 'frontier'. In 1766, de Bougainville observed Indians who come in bodies of two or three hundred men, to carry of [sic] the cattle from the Spaniards, or to attack the caravans of travellers. They plunder and murder, or carry them into slavery. This evil cannot be remedied; for, how is it possible to conquer a nomadic nation, in an immense uncultivated country, where it would be difficult even to find them?[7]

The struggle against the *indios bárbaros* (barbarian Indians) intensified following the Bourbon reorganization of the La Plata region into the Viceroyalty of La Plata in 1776, with Buenos Aires as its capital and administrative centre. In 1778, Spanish colonial officials ordered the second viceroy, Juan José de Vértiz y Salcedo, to establish settlements in Patagonia for the first time. The following year, the royal commissioner Francisco de Viedma y Nárvaez founded the fortress of Nuestra Señora del Carmen (subsequently known as Carmen de Patagones) at the mouth of the Río Negro—the southernmost Spanish outpost on the Atlantic coast. In 1790, Viedma's brother Antonio established a settlement further south at the Río Deseado estuary, which had to be abandoned when Viedma and his men came close to starvation. By 1783, the fortress at Carmen remained the sole outpost of civilization of Spanish governance in Patagonia, some 567 miles south of Buenos Aires. Further inland, the authority of Buenos Aires barely extended 110 miles to the Río Salado. Every year, expeditions of heavily armed men, ox-drawn wagons and horses trekked southwards to the enormous salt flats at Salinas Grandes, 500 miles south of the city, in order to collect the salt on which the dried beef industry depended.

By the end of the eighteenth century, the authorities in Buenos Aires had settled on a policy of containment against the raiding tribes of the south, based on a chain of forts, manned by the frontier militia known as Blandengues. These frontier militias consisted mostly of gauchos, on little or no pay, with no formal military training and no consolation for the heat and cold, the constant risk of Indian raids, and only the chapel, the occasional wife or *fortinera* (barracks woman) and the country bars known as *pulperías* to offer any respite from the privations of frontier life. The Bourbon authorities were painfully aware of the inadequacy of this system. In 1776, the new viceroy Pedro de Cevallos drew up a plan to raise an army of 10,000 to 12,000 men to drive the Indians out of the Pampas. Four years later, his successor, Juan José de Vértiz de Salcedo, appointed a commission to consider this proposal, which gloomily concluded that the 'peripatetic groups' of Indians in the south were unconquerable:

Their diet consists of mares and other animals distinct from those that we use; they don't need fire to eat, nor other provisions for their marches; they reside in mountains and uncultivated deserts; they travel through swampy, sterile and arid road; their robustness, accustomed to harsh conditions, endures to the point where we cannot even begin ... it is impossible for us to raise ten to twelve thousand men ... without their abandoning the arts and the agriculture to more misery than the benefit that would result from removing them.[8]

* * *

Unable to inflict a decisive defeat on the caciques of the south, the authorities in Buenos Aires combined occasional punitive expeditions with 'tributes' of animals and other gifts in exchange for peace and access to the salt flats. In 1796, the Spanish soldier, explorer and naturalist Félix de Azara (1746–1821) was commissioned to reconnoitre the south of Buenos Aires Province. Azara exhorted the government to attract more settlers to the south with grants of land in order to 'make the whites masters of the Pampa' and eliminate the threat of 'a few annoying barbarians'.[9] The obstacles to Azara's proposals will be immediately obvious to any contemporary visitor to the Pampas. Today, the population of Argentina is around 45 million, of whom around 15 million are concentrated in greater Buenos Aires. The remainder are spread across a country eleven times larger than the United Kingdom. In the early nineteenth century, the total population of the Argentine territories was around 450,000, of whom some 100,000 were enslaved Africans and their descendants. These figures did not include the *indios no sometidos* (unconquered Indians) further south. Falkner described 'two nations of Picunches and Pehuenches' who lived to the east and west of the Andes beneath the Chilean capital Santiago and the Argentinian city of Mendoza, who were 'formerly very numerous' but whose numbers were 'much diminished' through their wars with the Spanish as well as homicides and smallpox, which 'causes a more terrible destruction among them than the plague, desolating whole towns by its malignant effects'.[10]

In 1825, the British traveller John Augustus Barber Beaumont claimed that '[t]he present number of the Pampas Indians is computed at about eight thousand; they were formerly much more numerous'.[11] Neither Falkner nor Beaumont gives any sources for these figures, and the numbers of 'barbarians' who lived beyond the colonial frontier were easily exaggerated. Twentieth-century analysts of the Argentinian national character have frequently described Buenos Aires as a city looking towards Europe and haunted by its vast, unknowable interior. In the colonial era, these fears were frequently focused on the *indios bárbaros* of the Pampas, who could ride hundreds of miles across an unfenced wilderness to raid a town, fort or settlement before vanishing into the emptiness from which they came.

There were also more peaceful interactions between whites and Indians. By the early nineteenth century, Tehuelches from the far south of Chilean Patagonia were travelling regularly to Fort Carmen on the Atlantic coast to exchange feather dusters, ponchos, baskets, whips, lassoes, wooden stirrups, colt-leg boots and other Indigenous products for creole goods that included yerba maté (a herbal tea from the yerba maté plant), cloth, brandy, sugar, knives and raisins. These trading routes also extended to Buenos Aires itself. In 1820, the British artist and naval officer Emeric Essex Vidal observed Indians from the Pampas and 'the southern parts of Patagonia' in the capital, whose 'industry far surpasses that of the descendants of the Spaniards; for to the disgrace of the creole of this country, he is indebted to the savage Indian for the supply of many of his wants, and some of his luxuries'.[12] As in the United States, *indios amigos* (friendly Indians) or *indios mansos* (tame or domesticated Indians) worked on ranches or served as auxiliaries and *baquianos* (scouts) for colonial armies and rural militias. In 1802, the military commander of Buenos Aires Province described the Pampean Indians as 'very peaceful and there has been no incident for five years in which they have committed the least excess'. In June 1806, a British expeditionary force under the command of Sir William Carr Beresford occupied Buenos Aires and forced the viceroy to abandon the city. In his absence, the *cabildo* (town council) of Buenos Aires solicited help from two Tehuelche chieftains who offered to provide horses and auxiliaries to help drive the British 'Colorados'

(Redcoats) into the interior. The city fathers politely turned down this patriotic proposal and invited their 'faithful brothers' to watch the coasts and report on the arrival of British reinforcements instead.

* * *

These offers nevertheless contributed to what the late anthropologist and author Carlos Martínez Sarasola (1949–2018) called the 'Indigenist fever' that followed Argentina's 'May Revolution' and the establishment of its first independent government in 1810.[13] In January 1811, the ruling revolutionary junta decreed that Indigenous Peoples should have their own representative in each of the postcolonial administrative intendencies, and the *Gaceta de Buenos Aires* declared emphatically that 'the Indian is a citizen and falls under the protection of the laws'. That same year, a Tehuelche cacique and his entourage visited Buenos Aires, where the revolutionary leader Colonel Feliciano Chiclana hailed them as 'shoots from the same trunk … Friends, compatriots, and brothers.' On 12 March 1813, the General Assembly of Buenos Aires expressed its desire that 'the said Indians of the United Provinces be perfectly free men, and in equality of rights to all the other citizens that populate them'.

In 1816, at the Congress of Tucumán, the Argentine provinces established a provisional government, with Buenos Aires as the capital of the United Provinces of the Río de la Plata—the first iteration of what is now Argentina. Flushed with the universalist spirit of the French Revolution, the congress proposed to extend full political and civil rights to all the Indian inhabitants of the United Provinces. Anxious at their ability to persuade a largely monarchic Europe to accept independence, its representatives even briefly considered appointing an Inca king with a capital in Cuzco, Peru, to rule over the new state. In his famous 'general order' to the Army of the Andes on 27 July 1819 warning of a new Spanish expedition against the United Provinces, General Juan José de San Martín exhorted his troops to fight on even if they were forced to 'go naked like our compatriots the Indians'. In the years that followed independence in 1818, these expressions of solidarity gave way to a more enduring hostility, which increasingly regarded

31

Indians as the inveterate enemies of the incipient Argentine nation. For much of the first half of the century, the United Provinces were wracked by civil wars and power struggles between the 'unitarian' proponents of a single unified state with Buenos Aires as its capital and the 'federalists' who favoured provincial autonomy and self-governance.

Even in the midst of political chaos, the more intrepid ranchers and settlers continued to seek new lands in the south. In 1810, two Englishmen established the first *saladero*—a combination of slaughterhouse and salting plant—in the city of Buenos Aires, and this technical innovation was quickly adapted by ranchers outside the city. In 1824, the minister of government and foreign affairs of the United Provinces, Bernardino Rivadavia (1780–1845), negotiated a huge loan of 15 million pesos from the Barings Brothers Bank in London—the first foreign loan in Argentina's history—with the help of British intermediaries in the Argentine capital. As collateral for the loan, Rivadavia passed a 'law of emphyteusis', by which the state retained ownership of all public lands in Argentina while also allowing them to be rented out to its citizens.

In theory, this law was intended to encourage small tenant farmers to settle in the interior. In practice, some 21 million acres of public land were transferred to 500 wealthy landowners. In 1828, the Buenos Aires government extended its 'nominal frontier' southwards from Bahía Blanca on the southern Atlantic coast to Laguna Blanca Melincué to the west, effectively adding 75,000 square miles to the province. These developments intensified the competition for grazing lands, horses and cattle between Indians and settlers. On 3 December 1820, more than 2,000 Ranquel and 'Boroano' Indians—Mapuches from the district of Boroa in Chilean Araucanía—attacked the town of Salto, only 224 miles south-west of Buenos Aires. The raiders massacred the garrison, pillaged and burned the town and made off with 300 women and children and thousands of head of cattle. This *malón* (raid) was led by a dissident Chilean general-cum-outlaw named José Miguel Carrera, who had turned against the independent rulers of his own country and the United Provinces.

The following year, the governor of Buenos Aires, General Martín Rodríguez, sent an expedition in pursuit of the 'cursed monster that Chile vomited up', which finally captured Carrera in the town of Mendoza. Though Carrera claimed the raid had been carried out against his will, Rodríguez laid the blame squarely on 'that depraved man, that evil genius, that fury yawned by hell itself ... Barbarian, a hundred times barbarian and feline' who had incited 'the savages of the south' to attack Salto. In September 1821, Carrera was shot in the main square of Mendoza and his head exhibited on a pike. The 'malón de Salto'—Salto raid—was one of many 'unhappy civil dissensions', as the British diplomat and first Consul-General Sir Woodbine Parish (1786–1882) called them, that plagued post-independence Argentina, in which Indians fought alongside 'vagabond gauchos, deserters from the army, and such wretches flying from the pursuit of justice as, in times of civil commotion especially, are to be found in all countries'. In these confrontations, Parish wrote: 'Like bloodhounds it was impossible to restrain them. When once the weakest points were shown them, they burst in upon the frontier villages, murdering in cold blood the defenceless and unprepared inhabitants, and carrying off the women and children into a slavery of the most horrible description.'[14]

The English soldier, explorer and future lieutenant governor of Upper Canada, Sir Francis Bond Head, made various journeys across the Pampas in the 1820s, where he and his party were armed with loaded pistols to defend themselves against *salteadores* (highwaymen) and Indian raiders. Bond Head lamented the cruelty of these raids, in which men were tortured and mutilated and the more attractive women carried into captivity 'over the trackless regions before them, fed upon mare's flesh, sleeping on the ground, until they have arrived in the Indians' territory, when they have instantly to adopt the wild life of their captors'.

Bond Head described the 'savage, inveterate, furious hatred which exists between the Gauchos and the Indians' and recalled a 'very fine-looking Gaucho' who told him how he and his men had cut the throats of Indian prisoners captured after a battle—a fate that, according to Bond Head, 'the Indian firmly expects'.[15] Beaumont described the warfare on the Pampas as

inglorious and cruel; on the part of the Indians, it has consisted in driving away the cattle from the farms on the frontier; killing all

33

the men they could find, and carrying off the women and children. The retaliation has consisted in hunting them over the plains, and in like manner putting all the men to death; and bringing in the [Indian] women and children to Buenos Ayres, where they are made slaves.

Beaumont witnessed the arrival of nearly 200 Indian women prisoners in Buenos Aires, whose 'men had been put to death' and who were destined for domestic servitude. He criticized the rulers of Buenos Aires, who 'profess so strong a desire to increase the population of their country' that they had embraced 'the design of driving from the lands of their inheritance, or to exterminate the aboriginal inhabitants' even though the latter 'have given ample evidence of the docility of their nature, and of their aptitude to become excellent artisans and faithful troops'.[16]

* * *

Generals, politicians and ranchers might have dreamed of 'exterminating' the Indians of the Pampas, but they lacked the resources to realize even more modest military objectives. In March 1822, Colonel Pedro Andrés García was sent to the Salinas Grandes salt flats with orders to make peace with the Indian tribes of the south. Accompanied only by a small escort, García attended a marathon five-hour 'parliament', in which he tried in vain to persuade the Tehuelche chief Negro and hundreds of his warriors to accept a new frontier line south of the Río Salado. García expressed his frustration at his government's inability to hunt the Indians 'in their lairs, and make them pay for the many things they have taken from us, and rescue the slaves whom they have usurped from our industrious population'. He predicted a time when 'our resources prosper: then they will feel the weight of our vengeance, and a new era will begin, different from the one in which they took so much pleasure in assaulting us with impunity'.[17]

In the early nineteenth century, that moment was still far-off. In October 1823, 5,000 Ranquels and Tehuelches launched simultaneous attacks across a wide arc of territory reaching from the south of the province of Santa Fe to the south of Buenos Aires Province. In January the following year, Governor Rodríguez sent another 3,000-strong expedition to the south, equipped with seven pieces of artillery, with the intention of establishing a fort at the Río Negro—the northern boundary of Patagonia. Demoralized by cold, frostbite and lack of water, and repeatedly attacked by bands of Tehuelche warriors, the expedition was forced to withdraw before reaching the river. At this point, the 'Indian question' was a problem for which Argentina had no obvious solution. Of all the obstacles to 'the increase of a rapid population' in the interior, observed the Buenos Aires English-language newspaper the *British Packet* in 1827, there was

> none more powerful than the dread excited by the devastating incursions of the Indians, who behold the encroachments of civilization with a jealous eye, and whose legislative code has never taught them to respect either proprietorial rights, or from shedding innocent blood. They have looked upon these plains as their patrimony, and the fruits of others' industry as a lawful prey, and the peaceful settler as a usurper, with whom no terms could be observed short of either pillage, captivity, or death.

These 'peregrinating savages' had 'often been severely chastised by the troops of the Republic', the newspaper declared, but such punishments 'have had no more effect in mending their manners than birch-rod whippings ordinarily produce on truant-going schoolboys'.

One of the few men to have any significant impact on the Indians of the south was the Prussian army officer Friedrich Rauch (1790–1829), whom President Rivadavia commissioned in 1826 to 'eliminate the Ranquels' from the Pampas. A veteran of Napoleon's army in the Peninsular War and the Army of the Andes, Rauch waged pitiless war on Indians he regarded as heathen 'anarchists' and once boasted in his diary that his soldiers had 'cut the throats of 28 Ranquels' to save bullets. On 31 August 1826, Rauch's 350-strong cavalry regiment was forced to retreat following a violent clash with 800 Indians. The following day, Rauch's forces attacked *tolderías* that had nothing to do with this battle, killing some 200 Indians and Spanish and Chilean army deserters who were living with them. In October that year, Rauch's troops attacked Indian camps in the Sierra de la Ventana, some 350 miles south of Buenos Aires, killing

200 *indíos de lanza* (warriors) and taking 150 female prisoners, 12,000 head of cattle, 400 horses and sixty captives.

In 1828, the 'guardian of the frontiers' was presented with a sword from Rivadavia in recognition of 'the honour with which he has used his own in upholding the public cause'. In 1829, Rauch declared his loyalty to the unitarian general Juan Lavalle, who became governor of Buenos Aires after the execution of his predecessor, the federalist general Manuel Dorrego. As was often the case in Argentina's civil wars, Indians fought on both sides. On 8 March 1829, Rauch led a detachment of cavalry into battle against federalist forces and their Ranquel allies in a place called Las Vizcacheras, some 93 miles south of Buenos Aires. Some of these Ranquels had been on the receiving end of Rauch's 1826 operations. At one point in the battle, Rauch rode directly into the federal ranks, and a group of Ranquels brought his horse down with a bola. A Ranquel chief named Nicasio Maciel, known as 'Arbolito' ('Little Tree'), then administered the *coup de grâce* with a lance and cut off his head as a trophy. With Rauch's death, the United Provinces lost their most effective military commander. But by this time, the outside world had begun to take an interest in the territories that the governments of Argentina and Chile had largely ignored. In 1826, the British government sent two ships, the *Beagle* and the *Adventure*, to survey the channels and coasts along the southern tip of South America. And in 1831, the *Beagle* set out on its second voyage to the same territories, carrying the young naturalist and would-be clergyman whose writings would shortly transform Patagonia from a land of fable into a land of science.

3

SAVAGES, PRIMITIVES AND CANNIBALS

Even the most world-shaking historic events are often dependent on chance and coincidence. Had Captain Stokes, the commander of HMS *Beagle* during its first voyage of 1826–30, not shot himself in Tierra del Fuego in 1828, then Flag-Lieutenant FitzRoy would not have been promoted to captain when the *Beagle* docked in Montevideo for repairs. Had FitzRoy not been haunted by this suicide and his own family's history of depressive mental illness, he might not have sought a 'philosophic-companion' to help him ward off the 'blue evils' during the *Beagle*'s second voyage. Had the Reverend Leonard Jenyns not declined this invitation, then Darwin's Cambridge tutor John Henslow would not have recommended Darwin for the position. Had Darwin's uncle Josiah Wedgwood II not overcome his father's objections to this proposal, then the former college drop-out might have gone on to become a village parson, and the theory of evolution may have been developed by someone else.

The *Beagle*'s voyages took place at a time when Britain was just beginning to develop its 'informal' commercial and trading relationships with Latin America, in accordance with Foreign Secretary George Canning's famous boast to parliament in 1826 regarding Britain's recognition of the independent South American republics: 'I called the New World into existence to redress the balance of the Old.' In 1825, the British government signed a Treaty of Amity and Commerce with the United Provinces of the Río de la Plata. Even in these early stages, British influence on Argentina was such that the British consul Woodbine Parish noted that the typical gaucho was dependent on British manufactured goods, from 'the camp kettle in which he cooks his food' to 'the common earthenware he eats from—his knife, spoons, bits, and the poncho which covers him'.[1] By the time the *Beagle* set out on its second voyage, exports from the United Provinces were also largely transported in British ships. It was in this context that the *Beagle* was sent to take hydrographic and cartographical measurements of the Argentinian coast in 1826, in order to facilitate navigation between the Atlantic and the Pacific. In 1831, Captain Francis Beaufort (1774–1857), the British Admiralty Hydrographer of the Navy, commissioned FitzRoy to continue the hydrographic surveys of the South American coast south of the Río de la Plata that his predecessor had begun and also establish more accurate measurements of longitude through a circumnavigational voyage that would bring the *Beagle* back to England via Australasia and the Indian Ocean.

On the one hand, this voyage was intended to further Britain's commercial and naval interests in postcolonial South America through the charting of safe harbours, meteorological conditions, shipping routes, tidal currents. At the same time, the *Beagle* was a late participant in the tradition of sail exploration that had begun in the late seventeenth century, which combined national and commercial interests with the acquisition of scientific knowledge about the world and its peoples. Natural scientists, linguists, geologists and ethnographers all took part in these voyages, some of whom wrote bestselling accounts of their travels. Joseph Banks's account of Captain Cook's Pacific voyages; Louis Choris's *Voyage pittoresque autour du monde* (1822); Jean-François de Galaup de la Pérouse's *Voyage de la Pérouse autour du monde* (Perouse's voyage around the world, 1797); de Bougainville's 1767–8 Pacific journal; Alexander von Humboldt's *Personal Narrative of a Journey to the Equinoctial Regions of the New Continent* (1822)—these books caught the imagination of the wider European public, with their observations of flora and fauna, geology and geography, descriptions of exotic and faraway peoples, and lavish 'picturesque' images. An avid reader of Humboldt, Darwin was aware of the literary-scientific expectations that surrounded such expeditions. Like his illustrious predecessors, he intended to study flora, fauna and geology, collect fossils and specimens, and he also set out to write a Humboldtian 'personal narrative' of his travels. Even before leaving England, Darwin had a keen interest in a category of humanity that was already an established feature of European journeys to the 'unknown' regions of the earth.

* * *

The distinction between civilization and savagery is not an exclusively European construction; non-European societies have also divided humanity into similar categories. But long before the *Beagle* set out for South America, the world's 'savage' peoples had become a subject of particular interest to European scientists and the wider reading public. This fascination with the savage can be traced back to the 'monstrous races' with hooves and eyes in their chests that appear in the classical geographies of Pliny and Herodotus; in the construction of the 'barbarian' as an inferior category of humanity in classical Greece and Rome; in the hairy *Homo silvestris* or 'wild man' that appears in medieval and early modern European mythology; and also in the peoples that Europeans encountered—and sometimes invented—in their travels and explorations. In the fictionalized fourteenth-century travelogues purportedly written by Sir John Mandeville, the deserts surrounding the legendary Christian king Prester John's mythical African empire are inhabited by 'many wild men, that be hideous to look on; for they be horned, and they speak nought, but they grunt, as pigs'.[2] In the famous Valladolid debate of 1550–1, which considered the morality and justice of Spain's treatment of Indians in the New World, the theologian Juan Ginés de Sepúlveda (*c*.1489–1573) argued that 'the barbarians of the New World ... are as inferior to the Spaniards as children to adults and women to men' and that the difference between 'savage and cruel races and the most merciful, between the most intemperate and the moderate and temperate' was greater than 'between apes and men'.[3]

This was the basic template that Europeans applied to many different Indigenous Peoples: childlike, irrational and 'wild'; naked or nearly naked, devoid of language, laws, religion and property; prone to cruelty and cannibalism; and closer to animals than humans. To the French naturalist Georges-Louis Leclerc, Comte de

Buffon (1707–88), the 'savage nations' of his own era consisted of a 'tumultuous assemblage of barbarous and independent individuals' who were prone to cannibalism and infanticide and 'obey nothing but their own private interest, are incapable of pursuing one object, and submitting to settled usages, which supposes general designs, founded on reason, and approved of by the majority'.[4] There was also another tradition, extending from the sixteenth-century philosopher Michel de Montaigne to Jean-Jacques Rousseau, that imagined savage societies as an 'innocent' and 'unspoiled' counterpoint to a corrupt and vice-ridden civilization. By the late eighteenth century, these romantic iterations of the 'noble savage' were becoming rarer. The more Europeans ventured out into the world, the more they began to classify the 'varieties' or 'races' of mankind according to racial or cultural hierarchies that were defined by scientific or quasi-scientific criteria, in which the savage generally occupied the lower reaches of humanity.

In *De generis humani varietate nativa* (On the natural varieties of mankind, 1775) the comparative anatomist Johann Friedrich Blumenbach (1752–1840) identified five distinct human races: the Mongolian, Malayan, Ethiopian, American and 'Caucasian'—a category that referred to white Europeans. Blumenbach did not place these categories in a hierarchical order, but others did. To the English surgeon Charles White (1728–1813), Africans occupied 'the lowest degree of the human race', whereas the 'white European ... being most removed from the brute creation, may on that account, be considered as the most beautiful of the human race. No one will doubt his superiority in intellectual powers: and I believe it will be found that his capacity is naturally superior also to any other man.'[5] The French zoologist Georges Cuvier (1769–1832) contrasted the 'beauty' of the 'white, or Caucasic' race with its 'oval head form' to the 'yellow, or Mongolic' and 'negro, or Ethiopic' face whose 'projecting snout and thick lips put it visibly close to the apes'.[6] In his multi-volume *Systema naturae*, the Swedish naturalist Carl Linnaeus (1707–78) defined four varieties of humankind corresponding to Europe, Americas, Asia and Africa on the basis of the colour and texture of their hair, their temperament, their form of government and their clothing. In Linnaeus's system, *Europaeus albus* were 'white, sanguine, muscular', 'light, wise inventors' and 'protected by tight clothing', whereas *Africanus niger* were '[b]lack, phlegmatic, lazy' and 'sly, sluggish, neglectful'.[7] In the tenth edition, published in 1758, Linnaeus introduced the term *Homo sapiens* and added the two new categories of *Homo ferus* (wild children and youngsters) and *Homo monstrosus*—a category that included 'Alpine dwarfs', 'Patagonian giants' and 'Hottentots'.

If Caucasians or Europeans were generally associated with positive qualities such as physical beauty and intelligence, the First Peoples of the Americas often occupied the lower echelons of these hierarchical sequences. In 1771, the Dutch philosopher de Pauw described American Indians as 'idiot children, incurably lazy and incapable of any mental progress whatsoever'. That de Pauw had never been to the New World did not prevent him from describing its Indigenous inhabitants as 'people, who, it should seem, can never emerge out of infancy, or a state of nature'—a condition he attributed to the 'vitiated qualities' and 'malignity' of the American climate, which he believed led even Europeans to stagnation and degeneration.[8]

To the German racialist philosopher Christoph Meiners (1747–1810), the peoples of the Americas were 'unquestionably the most depraved among all human, or human-like creatures of the whole earth, and they are not only much weaker than the Negroes, but also much more inflexible, harder, and lacking in feelings'.[9] 'The temperament of the Indians is almost wholly undeveloped, and appears as phlegm', wrote the German botanist Carl Friedrich Philipp von Martius of his 1817–20 travels to Brazil. 'All the powers of the soul, even the more refined pleasures of the senses, seem to be in a state of lethargy. Without reflection on the whole of the creation, or the causes and internal connection of things, they live with their faculties directed only to self-preservation.'[10]

Many European travellers reached similar conclusions. The English translator's preface to Jacques-Julien Houtou de Labillardière's *Voyage in Search of La Pérouse* (1800) praised 'the laudable taste for voyages and Travels, which prevails in the present age', and provided opportunities for naturalists, geographers, merchants and also for the 'Moral Philosopher' who 'loves to trace the advances of his species through its various gradations from savage to civilized life' in order to 'deduce his conclusions respecting the social, intellectual, and moral progress of Man'. The translator reminded his readers: 'There appears no ground whatever for supposing, that any one tribe of mankind is naturally of an order superior to the rest, or has any shadow of right to infringe, far less to abrogate, the common claims of humanity' and exhorted philosophers to remember that 'the savage state of the most civilized nations now in Europe, is a subject within the pale of authentic history'.[11]

Such advice was not always heeded as information poured into Europe from the Enlightenment's voyages of exploration and discovery, all of which seemed to confirm the superiority of European civilization over the world's savage peoples. In an age of reason that increasingly defined itself as an age of science, the presence of Indigenous Peoples became a mystery to be solved, one that might shed light on the origins of humanity and enable scientists to trace mankind's 'various gradations from savage to civilized life'. For all these reasons, Azara described 'the Indian, the more savage he may be', as 'the principal and most interesting part of America'.[12] And in the early nineteenth century, few Indians were more interesting to Europeans than the inhabitants of the southernmost tip of the continent.

* * *

The peoples known collectively to Europeans as 'Fuegians' had already acquired a dismal reputation long before Darwin set eyes on them. In 1769, Captain James Cook called the Haush 'a little ugly half-starved beardless Race'. Returning from his second voyage to the Pacific in 1774, he described the Kawésqar people he mistakenly called 'Pecheras' as 'the most wretched' people he had seen, in whom 'one sees nothing about them that is not disgusting in the highest degree'. The German geographer Georg Forster (1754–94), who accompanied Cook on the same voyage, delivered an even more damning description of the 'Fuegians':

> They had little brown eyes, without life; their hair was black and lank, hanging about their heads in disorder, and besmeared with train-oil. On the chin they had a few straggling short hairs instead of a beard, and from their nose was a constant discharge of mucus

into their ugly, open mouth. The whole assemblage of their features formed the most loathsome picture of misery and wretchedness to which human nature can possibly be reduced.[13]

In Forster's opinion, the Fuegians' wretchedness was a confirmation of the 'pre-eminence of a civilized life over that of the savage'. 'Till it can be proved, that a man in continual pain, from the rigour of climate, is happy,' he wrote, 'I shall not give credit to the eloquence of philosophers, who have either had no opportunity of contemplating human nature under all its modifications, or who have not felt what they have seen.' Other sailors and explorers were equally contemptuous and condemnatory. In 1789, the Spanish naval officers Alejandro Malaspina and José Bustamante y Guerra compared the 'honesty' of the 'Patagonians' of the Argentine mainland to the Fuegians, whom they believed had violated 'the dictates of Nature itself' through their 'stupidity' in continuing to live in a territory that was wholly unfit for human habitation.[14] Captain Phillip Parker King, the commander of HMS *Adventure*, which accompanied the *Beagle*'s first voyage, also described the Fuegians as 'a most miserable, squalid race, very inferior, in every respect, to the Patagonians', who exuded an 'indifference, and total want of curiosity', which 'gave us no favourable opinion of their character as ethical beings; indeed, they appeared to be very little removed from brutes; but our subsequent knowledge of them has convinced us that they are not usually deficient in intellect'.[15]

This 'subsequent knowledge' derived almost entirely from the series of events that followed the theft of a whaleboat during the *Beagle*'s anchorage at Cape Desolation in the Strait of Magellan in January 1830 under its new commander, FitzRoy. When FitzRoy sent out a search party to retrieve the boat, the search provoked a violent altercation with a group of Fuegians, one of whom was shot dead. To prevent further incidents, FitzRoy took a group of hostages. Most of them escaped, apart from an eight-year-old girl called Yokcushlu, to whom FitzRoy's crew gave the name Fuegia Basket. FitzRoy subsequently acquired two male native captives—a twenty-eight-year-old man named El'Leparu, who was dubbed York Minster, and another young man whom they called Boat Memory. In May that year, FitzRoy took another hostage called

43

Orundellico, whom the crew renamed Jemmy Button, supposedly because his parents sold him in exchange for a mother-of-pearl button. FitzRoy believed that his captive belonged to the 'Tekenika' tribe—a misinterpretation of the Yagán word 'Teke unika', meaning 'I don't understand you.'

Button's 'exchange' was almost certainly due to a similar misunderstanding. FitzRoy and his crew did not speak any of the local languages, and it is unlikely that either 'Jemmy Button' or his parents understood what this transaction entailed. At first, FitzRoy was not certain what to do with the young man he called a 'displeasing specimen of uncivilized human nature'; his view of the Fuegians as 'satires on mankind' was very much in keeping with the disgust expressed by so many of his predecessors. He finally decided to take his captives back to England—an initiative that appears to have been motivated by altruistic and strategic considerations. FitzRoy later recounted his ambivalent feelings that led him to bring 'these ignorant, though by no means contemptible human beings' back to England. As 'disagreeable, indeed painful, as is even the mental contemplation of a savage, and unwilling as we may be to consider ourselves even remotely descended from human beings in such a state,' he wrote, 'the reflection that Caesar found the Briton painted and clothed in skins, like these Fuegians, cannot fail to augment an interest excited by their childish ignorance of matters familiar to civilized man, and by their healthy, independent state of existence'.[16]

Whatever their origins, FitzRoy told the Admiralty that the Fuegians might 'become useful as interpreters, and be the means of establishing a friendly disposition towards Englishmen on the part of their countrymen, if not a regular intercourse with them'. Following the Beagle's return to England in October 1830 from its first voyage, the Morning Post described the hopes of its captain that 'the condition of the savage inhabiting the Fuegian Archipelago may be in some measure improved, and that they may rendered less hostile to strangers. At present they are the lowest of mankind, and, without a doubt, cannibals.'

FitzRoy was very protective of his Fuegian captives and had them vaccinated against smallpox in Montevideo in order to protect them physically. Despite these efforts, Boat Memory died of smallpox at Plymouth Naval Hospital on 11 November that year. FitzRoy lamented the death of 'a great favourite with all who knew him, as well as with myself. He had a good disposition, very good qualities, and though born a savage, had a pleasing, intelligent appearance.'[17] The remaining three Fuegians were educated at an infant school in Walthamstow, where they were taught English, given religious instruction and shown how to eat from a table with knives and forks. The three Fuegians were also presented at court, where Queen Adelaide was so enchanted by Fuegia Basket that she gifted her a bonnet and a ring.

History does not record what these three young people felt about the society they had been brought to, but we know a great deal about what other people thought of them. On FitzRoy's orders, the Beagle's surgeon John Wilson conducted anthropometric and phrenological measurements of the Fuegians following their arrival in England, which found that 'the Fuegian, like a Cetaceous animal which circulates red blood in a cold medium, has in his covering an admirable non-conductor of heat' that enabled them to survive in cold temperatures. Wilson's skull measurements also revealed that '[d]estructiveness, secretiveness and cautiousness'— all faculties 'necessary to a savage warrior'—were 'large', whereas 'the more refined sentiments, [such] as benevolence, ideality, and conscientiousness' were 'small', along with 'nearly all the intellectual organs'.[18]

* * *

The Admiralty was lukewarm about FitzRoy's proposal to educate his four charges and return them to Tierra del Fuego as interpreters for future surveying missions. FitzRoy had therefore begun to make his own plans to return the Fuegians to their homeland out of his own funds, when he was commissioned by Beaufort to command another surveying mission in 1831. Reverend Henry Wilson, the Fuegians' guardian and teacher in Walthamstow, suggested to FitzRoy that he take two missionaries with him with a view to guiding the Fuegians towards a 'gradual civilization' and establishing a permanent Christian mission in their homeland. The Church Missionary Society agreed to send a young trainee missionary named Richard Matthews to Tierra del Fuego for this purpose. This experiment inevitably piqued the interest of the 'gentleman-companion' he had recruited to help him stave off the 'blue evils'. In a diary entry on 13 November, Darwin records his first meeting with the three Fuegians at Plymouth, accompanied by Reverend Wilson and 'Matthews the missionary'.[19]

At this point, these Fuegians are still the sweet Anglicized natives in short haircuts, polished faces and English clothes whom FitzRoy sketched alongside the unkempt, animal-like renditions of their original 'savage state' in order to demonstrate how their 'coarse' features, as FitzRoy put it, were 'much improved by altered habits, and by education' during their stay in England. Darwin frequently conversed with the three Fuegians during the *Beagle*'s Atlantic crossing and observed their interactions with the crew and with each other. York Minister was 'reserved, taciturn, morose, and when excited violently passionate'—a moodiness that was partly due to his attraction to Fuegia Basket, whom Darwin describes as 'a nice, modest, reserved young girl, with a rather pleasing but sometimes sullen expression, and very quick at learning anything, especially languages'. Jemmy Button is 'the universal favourite'—a vain dandy who is fond of admiring himself in a mirror and who also sympathetically exclaims 'Poor fellow! Poor, poor fellow!' during Darwin's frequent bouts of seasickness.[20]

Darwin comments on the Fuegians' acute eyesight—a quality he considered to be more common to 'men in a savage state, as compared with those long civilized'. He also complains that 'it was singularly difficult to obtain much information from them, concerning the habits of their countrymen', due to their childlike inability to consider alternative answers to his questions. In these early observations, the Fuegians are still recognizable as individuals, with their distinctive personalities. If they have been savages before, they still carry enough traces from their years in England for Darwin to recognize them as semi-civilized members of the same species—however enigmatic and impenetrable. But his attitude towards their countrymen was a different matter. In a letter to a university friend named Charles Thomas Whitley in September 1831, Darwin joked that he was buying a set of pistols for 'fighting with those d- Cannibals' and 'the King of the Cannibals Islands' whom he expected to encounter in his travels.[21] In effect, 'savages' were part of Darwin's itinerary even before he left England, as an exotic and slightly alarming attraction and an essential object of intellectual curiosity and fascination for the *voyageur philosophe* (traveller-philosopher) who was seeking to understand the world and his place in it.

4

THE MAN ON THE ROCK

Darwin's first encounter with the Indigenous Peoples of the New World took place at Bahía Blanca, in southern Buenos Aires Province, where the government of the United Provinces had recently established a 'protective fortress' during the 1825–8 Cisplatine war between the United Provinces and Brazil. On 7 September 1832, the *Beagle* anchored in the bay, and Darwin rowed ashore with FitzRoy and the ship's purser George Rowlett. They were greeted by a welcoming party of 'wild Gaucho cavalry'. Darwin was instantly smitten with Argentina's emblematic horse-riding 'cowboys' and their exotic attire of ponchos, baggy trousers and the rectangular skirt-like breeches or *chiripá* wrapped around their waists. It was, the young naturalist wrote, 'the most savage picturesque group I ever beheld—I should have fancied myself in the middle of Turkey by their dresses'. The *Beagle* spent nearly two months at Bahía Blanca, where Darwin spent his time hunting, botanizing and digging for fossils in the cliff face at nearby Punta Alta with his assistant Syms Covington and other *Beagle* crew members. In the course of these excavations, Darwin uncovered the giant quadruped skull that the anatomist and palaeontologist Richard Owen later identified as a variant of the ground sloth. It was this discovery that first led Darwin to consider the natural processes through which variations of the same species became extinct and that also convinced him that the 'sterile country a little southward, near the Río Negro ... would support many and large quadrupeds'.[1]

These first investigations into the Pampean evolutionary necropolis were conducted in the midst of a brutal conflict between whites or 'Christians' and the Indians of the Pampas. The Bahía Blanca fortress was subject to frequent attacks by 'large bodies of Indians' in a war that Darwin learned was carried on 'in the most barbarous manner. The Indians torture all their prisoners & the Spanish shoot theirs.' Shortly before the *Beagle*'s arrival, an expedition from the fortress had attacked an Indian encampment and seized horses and prisoners. When two caciques attempted to negotiate their release, the 'Spaniards', as Darwin called them, shot them both. This was Darwin's first glimpse of settler-colonial warfare, and he was not enamoured with either side. At the protective fortress, he was disgusted by the *indios amigos* 'gnawing bones of beef', who made him think of 'half-recalled wild beasts. No painter ever imagined so wild a set of expressions.' During a hunting expedition with a party of gauchos, 'pure Indians' and 'others most ambiguous; but all alike were most wild in their appearance', he noted 'the swarthy but expressive countenances of my half-savage hosts'. To Darwin, both gauchos and Indians occupied an intermediate zone of racial and cultural ambiguity, between 'wildness' and 'civilization'. It was not until December that year that the *Beagle* returned to Jemmy Button's country, and Darwin finally had the opportunity to observe savages 'in their natural state' for the first time.

* * *

From Bahía Blanca, the *Beagle* sailed north to Montevideo and Buenos Aires. In Montevideo, Darwin attended a Rossini opera and a grand ball and enjoyed dinner and 'delightful shady walks' at the estancia of an expatriate Englishman of his own class. In Buenos Aires, he visited a museum and a church and browsed the 'many shops kept by Englishmen & full of English goods'. He also took time out to ogle Argentinian women. 'Our chief amusement was riding about & admiring the Spanish Ladies,' he told his sister Caroline. 'After watching one of these angels gliding down the streets; involuntarily we groaned out, "how foolish English women are, they can neither walk nor dress". And then how ugly Miss sounds after Signorita; I am sorry for you all.'[2] On 26 November 1832, the *Beagle* headed south once again. And the following month, Darwin encountered the Haush 'devils' at Good Success Bay, whom he compared in a letter to his tutor Henslow to 'troubled spirits from another world'.[3] In less than a month, Darwin had left a world of balls, English shops, opera and attractive 'Spanish Ladies' and found himself among people who, in his eyes, were only barely recognizable as members of the same species. 'Their language does not deserve to be called articulate,' he wrote in his diary. 'Capt. Cook says it is like a man clearing his throat ... Imagine these sounds & a few gutterals mingled with them, & there will be as near an approximation to their language as any European may expect to obtain.'

To Darwin, the absolute savagery of the Haush was evident in their 'miserable' dress and appearance, their food and the absence of property 'excepting bows & arrows & spears'. Their 'very attitudes were abject, & the expression distrustful, surprised & startled'. These 'attitudes' softened when the crew distributed pieces of red cloth as gifts, after which, Darwin observed ironically, 'we became good friends'. It was not a relationship between equals. Like Captain Cook before him, Darwin concluded that '[i]f the world was searched, no lower grade of man could be found. The Southsea Islanders are civilized compared to them, & the Esquimaux, in subterranean huts may enjoy some of the comforts of life.'

These damning conclusions were an extension of Darwin's equally negative response to Tierra del Fuego's landscape and climate. 'Whilst looking round on this inhospitable region we could scarcely credit that man existed in it,' he wrote on Christmas Day. A connoisseur of 'sublime' landscapes, Darwin was not immune to the 'still solitude' of Tierra del Fuego's ancient forests, where 'death instead of life is the predominant spirit', but these fleeting attractions were soon eclipsed by cold, rain and physical discomfort. The following month, the *Beagle* struggled to make headway through the Beagle Channel in the face of gales and rough seas. All this intensified Darwin's chronic seasickness—and his antipathy to Tierra del Fuego and its inhabitants. On 20 January 1833, Darwin saw another group of 'savage & wild' Fuegians running along the shore, some of whom had lit fires to announce their arrival. 'Four or five men suddenly appeared on a cliff near to us,' he wrote. 'They were absolutely naked & with long streaming hair; springing from the ground & waving their arms around their heads, they sent forth most hideous yells. Their appearance was so strange, that it was scarcely that of earthly inhabitants.'

These men were Yagáns—Jemmy Button's people. That same day, the crew went ashore to meet them, and Darwin heard the word 'Yammerschooner', which the natives used repeatedly. This word probably comes from the Yagán word *yamask-úna*—meaning 'be kind to me' or simply 'please', but Darwin and the *Beagle* crew took it to mean 'Give me', and it quickly became synonymous in his mind with childlike scrounging. On 22 January 1833, the *Beagle* anchored at Ponsonby Sound, north-west of Navarino Island, where members of Button's tribe told them that Button's father had died in his absence. 'We were sorry to find that Jemmy had quite forgotten his language,' Darwin recorded. 'It was pitiable, but laughable, to hear him talk to his brother in English & ask him in Spanish whether he understood it.' According to Darwin, Button had already dreamed of his father's death and therefore 'seemed to expect it & did not much care about it, though he comforted himself with the expression "me no help it"'. That Button needed to console himself suggests that he did care about his father's death, and Darwin's discovery that the Fuegians never 'mention the dead' provides another plausible explanation for his apparent lack of concern.

On 24 January, Button was finally reunited with his mother, brother and uncle at Wulaia Cove on Navarino Island, a meeting Darwin describes as 'not so interesting as that of two horses in a field'. FitzRoy used another animal analogy in a letter to his sister, declaring that '[s]trange dogs meeting in a street shew more anxiety and more admiration than was manifested at this inhuman meeting of a lost child and his afflicted mother and relatives'.[4] Both Darwin and FitzRoy seemed to have assumed too much, or too little, about this reunion, even though York Minster later told them that Button's mother had been 'inconsolable' after her son's disappearance and had searched Tierra del Fuego for weeks in the hope that the *Beagle* had left him behind.

Having returned FitzRoy's captives to their native land, the *Beagle* crew spent the next four days building 'wigwams' where Matthews and the Fuegians were expected to form the Church Missionary Society's permanent settlement. It is not known whether the four Fuegians actually wanted to live together, though a separate shelter was built for York Minster and Fuegia Basket, who were now considered to be 'married'. The crew also dug vegetable plots for the mission and equipped it with chamber pots, soup tureens, a tea set, crockery, dresses, shirts and white linen and other unlikely accoutrements of English civilization, which the London Missionary Society had acquired from charitable donations. On 28 January, Darwin, FitzRoy and some crew members left Matthews and the Fuegians and went off in boats to survey the islands to the west of the Beagle Channel.

On 6 February, they returned to find that groups of Fuegians had threatened Matthews and stolen most of the settlement's possessions. FitzRoy advised the terrified missionary to abandon the island. In little more than a week, FitzRoy's attempt to establish an outpost of England in Tierra del Fuego had collapsed—a failure Darwin attributed primarily to

> the perfect equality of all the inhabitants [which] will for many years prevent their civilization: even a shirt or other article of clothing is immediately torn to pieces. Until some chief rises, who by his power might be able to keep to himself such presents as animals &c &c, there must be an end to all hopes of bettering their condition.

Despite feeling 'quite melancholy leaving our Fuegians amongst their barbarous countrymen', Darwin still believed that 'in contradiction of what has often been stated, 3 years has been sufficient to change savages, into, as far as habits go, complete & voluntary Englishmen'. Even York Minster, with his 'strong, violent mind', would 'in every respect live as far as his means go, like an Englishman', whereas '[p]oor Jemmy, looked rather disconsolate, & certainly would have liked to have returned with us'.

. Though Darwin believed that the Fuegians' 'excursion to England' would not be 'conducive to their happiness', he believed that FitzRoy's captives 'have far too much sense not to see the vast superiority of civilized over uncivilized habits; & yet I am afraid to the latter they must return'. The following week, FitzRoy returned to the settlement and gave Darwin 'a very prosperous account' of it. The Fuegians had not been robbed, and 'the conduct of the natives was quite peacible [sic]', all of which suggested to Darwin that 'this little settlement may be yet the means of producing great good &

altering the habits of the truly savage inhabitants'. Darwin sounded less hopeful in his letter to his sister Caroline in March from the Falkland Islands that year. 'These Fuegians are Cannibals,' he wrote, 'but we have good reason to suppose it carried on to an extent which hitherto has been unheard of in the world.'[5]

These claims stemmed primarily from a local sealing boat captain named William Low, who told the crew that the Fuegians 'devour their old women before they kill their dogs'. According to Darwin, when Low asked his Fuegian cabin boy why they did this, he was told: 'Doggies catch otters, old women no.'[6] The missionary Matthews was also reportedly told by Button that his people 'sometimes eat the women' in periods of scarcity. This was not the most conclusive evidence, and it was to some extent contradicted by Darwin's own claim in his journal that Button 'will not eat land-birds because they live on dead men'—an aversion that does not suggest a taste for human flesh. Like many European travellers of the period, Darwin believed that 'savage' women were treated as beasts of burden, and from this point it was easy to assume that women—particularly old women—would be disposable and edible in times of crises. All these perceived customs filled him with 'a disgust at the very sound of these miserable savages', and yet in the same letter he told his sister that the 'extreme interest which is created by the first sight of savages' was something that 'almost repays one for a cruize in these latitudes; & this I assure you is saying a good deal'.

* * *

The *Beagle* spent the rest of the year surveying the Patagonian coastline further north and assisting in the establishment of a British settlement in the Falkland/Malvinas Islands and did not return to 'Jemmy Button's country' until February 1834. During those twelve months, Darwin had gained some idea of the range of Patagonia's Indigenous Peoples. At Carmen de Patagones—the successor to Fort Carmen—he saw 'tame' or domesticated Indians, who lived on government rations and traded with the local whites. Darwin had mixed feelings about this proximity. 'What their character may have gained by lessening their ferocity, is lost by their entire immorality,' he wrote in his journal. 'Some of the younger men are however improving; they are willing to labour … They were now enjoying the fruits of their Labour, by being dressed in very gay, clean clothes & being very idle.' At Bahía San Gregorio, in the Strait of Magellan, the *Beagle* crew bartered with 'half-civilised' Tehuelche Indians, who had learned Spanish and English from their contacts with sealers, three of whom were invited on board for tea, where they 'behaved quite like gentlemen, used a knife & fork & helped themselves with a spoon'. Darwin also met two Selk'nam 'foot Patagonians' with a 'facility for learning languages' that he predicted 'will greatly contribute to their civilization or demorilization [the loss of their own cultural integrity]: as these two steps seem to go hand in hand'.

These encounters reinforced his revulsion at the sight of the six Yagáns who approached the *Beagle* in a canoe after a year's absence. This was the famous scene that the *Beagle*'s second artist Conrad Martens painted in one of his most iconic watercolours, which shows a Fuegian standing in his canoe and waving at the *Beagle* in what appears to be a gesture of welcome. Though Martens shared Darwin's view of the Fuegians as 'the most destitute of human beings', he also concluded that 'poverty however is seldom injurous to picturesque effect, and these figures are all that can be desired in that respect'.[7] These effects are evident in his haunting depiction of the Yagáns and the *Beagle* as an encounter between two different worlds. Darwin's response was very different—'I never saw more miserable creatures,' he wrote on 25 February 1834:

> Stunted in their growth, their hideous faces bedaubed with white paint & quite naked. One full aged woman absolutely so, the rain & spray were dripping from her body; their red skins filthy & greasy, their hair entangled, their voices discordant, their gesticulation violent & without any dignity. Viewing such men, one can hardly make oneself believe that they are fellow creatures in the same world.

Once again, Darwin asked himself whether 'there is any spectacle more interesting & worthy of reflection, tha[n] one of these unbroken savages. It is a common subject of conjecture; what pleasure in life some of the less gifted animals can enjoy? How much more reasonably it may be asked with respect to these men.'

He listed the conditions that placed this 'pleasure in life' beyond
the reach of the Fuegians: the 'tempestuous climate', where they
'sleep on the wet ground, coiled up like animals'; their 'miserable
food'; their 'wretched' canoes; the proximity of 'hostile tribes
speaking different dialects' and the 'wild rocks, lofty hills & useless
forests ... viewed through mists & endless storms', where they had
established themselves. Trapped in this environment, Darwin saw
the Fuegians as tragic victims of their own cultural retardation,
'compelled unceasingly to wander from spot to spot' with their
attention entirely focused on survival. In these circumstances, '[h]
ow little can the highest powers of the mind come into play,' he
asked, 'what is there for imagination to paint, for reason to compare,
for judgment to decide upon'. Though he recognized the Fuegians
as 'essentially the same creature' as himself, he struggled to account
for the 'scale of improvement ... comprehended between the
faculties of a Fuegian savage & a Sir Isaac Newton'. This gulf raised
other questions:

> Whence have these people come? Have they remained in the same
> state since the creation of the world? What could have tempted a
> tribe of men leaving the fine regions of the North to travel down
> the Cordilleras the backbone of America, to invent & build canoes,
> & then to enter upon one of the most inhospitable countries in the
> world. Such & many other reflections, must occupy the mind of
> every one who views one of these poor Savages.

Darwin conceded that 'the race of Fuegians' was not decreasing,
and that therefore they must enjoy 'a sufficient share of happiness'
to make their environment worth living in. He also recognized
that '[n]ature, by making habit omnipotent, has fitted the Fuegian
to the climate & productions of his country'. But this grudging
appreciation of the habit of survival was very different from the
abilities that the anonymous narrator of the Argentinian novelist
Sylvia Iparraguirre's novel *Tierra del Fuego* (1999) ascribes to his
friend Jemmy Button:

> Button knew much more than I did about anything that came
> up. He was a better sailor, he had astonishing eyesight and even
> more astonishing marksmanship with stones; he could be naked
> in freezing rain or dive into an ice-cold sea; he knew how to hunt

and gather shellfish and find cormorant nests on the cliffs; he knew which kind of penguins were not good to eat and where to find fresh water and firewood. He had probably already been with a woman, and, within a year, if he were staying on land, he would be a father.[8]

It might seem unfair to compare Darwin's responses to the observations of a twentieth-century novelist, but Iparraguirre's insights were not unknown in Darwin's time. 'The inhabitants of Tierre del Fuego have also been spoken of as if they were beings possessed of little more than animal instinct, incapable of being instructed,' wrote the British explorer James Weddell:

> This may, perhaps, be the case; arising, however, out of the peculiar situation in which they are placed. Give them intercourse with foreigners, and they will improve with understanding; for I have found them to be not only tractable and inoffensive, but also, in many of their employments, active and ingenious.[9]

During the *Beagle*'s first voyage to Tierra del Fuego, Stokes considered the seal oil the Fuegians used to keep themselves warm to be 'miserably suited' to the weather conditions and more effective than European clothing. Stokes noted the Fuegians' 'affectionate and caressing' treatment of their children and the 'tenderness with which they tried to calm the alarms our presence at first occasioned, and the pleasure which they showed when bestowed upon the little ones any trifling trinkets'.[10]

In 1833, Royal Naval Officer John Macdouall, the clerk on the 1826–9 *Beagle* voyage, published an account of his travels in which he described 'several instances' in which the Fuegians 'appeared to be very affectionate to their children'. Macdouall was impressed by their canoe-building skills and the 'ingenuity and perseverance' with which they used only mussel shells as work tools. At Bahía San Gregorio, he observed in the Tehuelche 'a contentment and a delight in their native wilds inconceivable to the inhabitants of the civilised world'.[11] Such possibilities are entirely absent from Darwin's writings about the Fuegians or any other 'savages' who had not been exposed to civilization. In the published version of his journal, Darwin returned once again to the theme of Fuegian cannibalism and cited an anecdote by John Byron, one of the survivors of the shipwreck of the British warship HMS *Wager* in 1741. In his account of the shipwreck that was written some years afterwards, Byron describes how he and some of his shipmates were helped by a group of Indians on the western coast of Patagonia, led by a 'Christian cacique'. Instead of taking them to safety at the island of Chiloé, the Indians abandoned them, and the cacique beat his child to death against a rock in a fit of temper because the child dropped a basket of sea eggs.[12]

Byron was not the most reliable of narrators, and the Indian who carried out this alleged infanticide was not even a 'Fuegian' but a Christianized chief from the Patagonian mainland. Yet this story was accepted by Darwin as prima facie confirmation of Fuegian barbarity. 'Was a more horrid deed ever perpetrated,' he asked his readers, 'than that witnessed on the west coast by Byron, who saw a wretched mother pick up her bleeding dying infant-boy, whom her husband had mercilessly dashed on the stones for dropping a basket of sea-eggs!' According to Darwin's biographer Janet Browne, it was the 'experience of comparing Anglicizied Fuegians with native tribesmen' and 'native Fuegians with himself' that gave Darwin 'the breadth of vision to include mankind as an integral part of the natural world'.[13] But this breadth was often strikingly absent from Darwin's reductionist descriptions of the 'savages' he encountered in Tierra del Fuego. On 1 March 1834, Darwin described a group of Fuegians as 'very quiet & civil & more amusing than any Monkeys. Their constant employment was begging for everything they saw; by the eternal word—yammer-scooner.'

Darwin found this 'begging' tiresome and annoying, despite the opportunities it provided for 'looking & laughing at these curious creatures'. On 5 March, the *Beagle* returned to Wulaia, where Button came out in a canoe to meet them. 'It was quite painful to behold him,' Darwin wrote,

> thin, pale, & without a remnant of clothes, excepting a bit of blanket round his waist: his hair, hanging over his shoulders; & so ashamed of himself, he turned his back to the ship as the canoe approached. When he left us he was very fat, & so particular about his clothes, that he was always afraid of even dirtying his shoes; scarcely ever without gloves & his hair neatly cut. I never saw so complete & grievous a change.

Button was invited on board, where his hosts clothed and cleaned him up until 'things wore a good appearance'. He also gave two 'beautiful otter skins' to two of his friends on the ship, in addition to some spear heads and arrows he had made specially for FitzRoy. His hosts were pleased to learn that Button 'had got a young & very nice looking squaw [wife]' and 'had not the least wish to return to England', even though he had struggled to re-learn his own language. Button told his friends that York Minster had treated Fuegia Basket 'very ill' and persuaded Button and his mother to return to the Ona (Selk'nam) country in northern Tierra del Fuego, where he robbed and abandoned them.

The next day, Button came out to see his friends once again, and the meeting came to an end when Button's wife came on board and begged him to leave the boat, seemingly anxious that he might not return. 'Every soul on board was as sorry to shake hands with poor Jemmy for the last time, as we were glad to have seen him,' Darwin wrote. 'I hope & have little doubt he will be as happy as if he had never left his country.' As the *Beagle* left Ponsonby Sound, Button lit a farewell signal fire for his friends, and this was the last time Darwin saw him. Despite his disappointment at Button's reversion from the 'well-dressed stout lad we left him' to a 'naked, thin, squalid savage', as he told his sister Catherine, Darwin did not feel any regrets about leaving Tierra del Fuego.[14]

On 23 July that year, he wrote a letter from Valparaíso to Whitley—the old Cambridge friend to whom he had once joked about buying pistols to defend himself against cannibals. By this time, Darwin had been away from England for more than two and a half years, and his letter was steeped in nostalgia and homesickness. He quoted from Jane Austen and described the 'glimpses into futurity' based on 'retirement, green cottages & white petticoats' that seemed lost to him as a result of his participation in an endless voyage that deprived him of his 'friends & other sources of pleasure' in Cambridge. In an attempt to justify his voyage to Whitley, and perhaps to himself, he described the 'solid enjoyment, some present but more in anticipation' that he took from geology and from the 'glory & luxuriance' of the tropical scenery that 'exceeds even the language of Humboldt to describe'. He then turned to what was already a familiar theme in his correspondence:

But I have seen nothing, which more completely astonished me, than the first sight of a Savage; it was a naked Fuegian his long hair blowing about, his face besmeared with pain. There is in their countenances, an expression, which I believe to those who have not seen it, must be inconceivably wild. Standing on a rock he uttered tones & made gesticulations than which, the crys of domestic animals are far more intelligible. When I return to England you must take me in hand with respect to the fine arts ... How delightful it will be once again to see in the Fitzwilliam, Titian's Venus; how much more delightful to go to some good concert or fine opera.[15]

This juxtaposition of the 'naked Fuegian' with the civilized pleasures of Titian's *Venus*, concerts and 'fine opera' was on one hand a product of Darwin's nostalgia and homesickness. At the same time, the savage-on-the-rock was also a mark of pride—a validation and a demonstration of how far Darwin had travelled in space and time, from Jane Austen and petticoats to a primitive world whose inhabitants spoke a language that was less intelligible than the crying of domestic animals. This was the central image of 'the savage' Darwin presented to his correspondents and that remained with him for the rest of his life.

* * *

In the shrunken, interconnected and increasingly homogenized twenty-first century, where journeys that once took months or even years can be compressed into a few hours, it is difficult to imagine these temporal and physical distances. Today, it takes little more than an hour to fly from the Chilean city of Punta Arenas to the Isla Navarino, where Darwin and FitzRoy said goodbye to Button for the last time in 1834. Visitors with more time on their hands can take the splendidly named Transbordadora Austral (Southern Shuttle) ferry service from Punta Arenas, on the Strait of Magellan, which takes thirty-four hours to reach Puerto Williams, the main settlement on the Isla Navarino. When I sailed on board the *Yaghan*, my fellow-travellers included a Chilean astronomer and his social worker wife and a former hockey player in the Chilean national team who had come to the 'end of the world' to pay homage to his late best friend and fellow hockey player by setting fire to his former team shirt and floating it out to sea, to the soundtrack of Deep Purple's 'Smoke on the Water' on his mobile phone.

A picture of Darwin's venerable bearded face greeted us three times a day when we queued up for our meals, but much of our time was spent on deck, in a state of permanent wonder at the glaciers and snow-covered mountains, the labyrinthine channels and forest-covered archipelago that connect the Strait of Magellan to the Beagle Channel. This enchantment was generally absent from the writings of Europeans who navigated these waters in the age of sail. The weather was overcast and drizzly for much of our voyage, and the patches of open sea were still choppy enough in places to appreciate the skill and bravery of the European seamen who battled with adverse currents, storms and rain, and also of the First Peoples who navigated these waters for thousands of years in their bark canoes. In Patricio Guzmán's meditation on Patagonian history and geography, *The Pearl Button* (2015), an elderly Kawésqar man casually describes canoeing with his father to Cape Horn— the dreaded passageway for so many European sailors. Now these canoes are long gone, along with the wooden 'wigwams' the Kawésqar and the Yagáns erected as they canoed back and forth through their watery domain. Today, some seventy-four Yagáns live in the Villa Ukika, a neighbourhood of clapboard houses on the edge of Puerto Williams—the southernmost city in the world.

The designation of 'city' is a bold claim for a settlement of fewer than 5,000 inhabitants, most of whom are connected with the Chilean navy, on a mountainous island where the silence is broken only by the occasional vehicle. One morning, I watched Maurice van de Maele, the owner of our hotel, chase away the horses that wander freely through the streets, which were eating the flowers from his window-boxes. No sooner had the horses disappeared than a cow came charging past the window, pursued by a dog. This was the most dramatic event of the day. Before moving into the hotel business, Van de Maele was the director of the Martin Gusinde Anthropological Museum, just around the corner, where he worked for seventeen years. He described his affinity for cold remote places and his surprise and delight at his first sight of Tierra del Fuego's 'dense humid forests'.

Why did he think Darwin and so many others had seen that same landscape so differently? 'When you have a firm attachment to this place, it's full of fabulous ways to survive,' he replied:

> But the Europeans came from a long way away, and when looking at the landscape they didn't understand that there were animals here. They didn't know how to catch them ... For a Yagán living here, it's really easy to survive. With all the technology we have, with all our accumulated intelligence, we aren't capable of inhabiting a single island. In Puerto Williams we survive because boats and planes bring us many things, but there's no one on these islands. Seven thousand years ago they were filled with people. They were healthy and happy, and this culture could teach us many things.

Today, there is a vogue for survivalist reality television shows in which urban Westerners are abandoned naked or semi-naked on desert islands or left to survive and learn bushcraft skills in deserts and jungles. These shows are partly a form of voyeuristic entertainment, with their teary confessions-to-camera and character-building ordeals, but in a world where millions of people buy food in supermarkets with little idea where it comes from, there is an inevitable fascination in imagining how some of us might cope if these systems ever broke down—and a greater willingness to appreciate the skill and ingenuity of peoples who survived for thousands of years before these systems were constructed. Few nineteenth-century visitors were willing to concede that the people they regarded as savages might have shown courage, skill and ingenuity in surviving in an environment that Europeans found hostile and dangerous. Unable to speak a common language, or look beyond their own preconceptions, too many Europeans regarded the Fuegians as Darwin did—as objects of pity, revulsion and contempt. But for Darwin, the existence of such people touched on questions he was only just beginning to ask himself about how different species evolved and disappeared. When he saw the Yagáns for the last time, he could not have imagined that he was part of a chain of events that would ultimately destroy their society. But further north, he had already witnessed the first attempt by the government of Buenos Aires to conquer an Indigenous population it regarded as barbarian savages.

5

EXTERMINATION

In his travel journal and letters, and in his expeditions into the South American interior, Darwin was very conscious of the Enlightenment tradition of the *voyageur-philosophe* epitomized by his hero Humboldt. Like Humboldt, he embraced physical hardship and danger in pursuit of scientific knowledge. He went off the beaten track and combined scientific observations and reflections in his writings with rapturous descriptions of landscapes and dramatic events. A humane and gentle man who was haunted all his life by the 'most pitiable moans' of a slave he overheard being whipped in Pernambuco, during the *Beagle*'s second visit to Brazil in 1836, Darwin frequently recounts stories of murders and robberies, throat-cutting bandits and coups, massacres, earthquakes and revolutions with the macabre relish of an after-dinner raconteur.

During the *Beagle*'s first stopover at Bahía Blanca in September 1832, he witnessed the face-to-face warfare between white settlers and Indigenous Peoples while excavating the 'perfect catacomb for monsters of extinct races' at the cliffs in Punta Alta. It was during these excavations that Darwin first used a word that flits repeatedly through his writings when he observed that 'no fact in the long history of the world is so startling as the wide and repeated exterminations of its inhabitants'. In this case, 'extermination' has the passive meaning of extinction or disappearance rather than the active associations of the verb 'to exterminate'—meaning 'destroy completely'—and Darwin used it to refer to the 'monsters' whose fossils he had begun to uncover in the cliffs at Punta Alta. But the same term was frequently used in the nineteenth century in reference to the world's 'savage' peoples. 'Few future events are more certain than the speedy extermination of the Indians of N. America and the savages of New Holland', predicted Charles Lyell in the second volume of *Principles of Geology* (1832), which Darwin received in South America—an outcome Lyell called 'the faint image of the certain doom of a species less fitted to struggle with some new conditions'.[1] The concept of 'speedy extermination' also had the more specific meaning of physical removal or elimination in settler-colonial conflicts over land and resources. In an address to the California Senate and Assembly in 1851, the state governor Peter Hardeman Burnett called for 'a war of extermination ... between the races until the Indian race becomes extinct'. Though Burnett acknowledged that '[w]e cannot anticipate this result but with painful regret', he insisted that 'the inevitable destiny of the race is beyond the power or wisdom of man to avert'.[2]

In this case, extermination is both a 'destiny' and an active process—the annihilation of an entire people, and this was how the term was often understood in nineteenth-century settler-colonial discourse. It was not until the mid-twentieth century that the Polish jurist Raphael Lemkin coined the term 'genocide' to describe the mass killings carried out by the Nazis in Occupied Europe. But the physical destruction of Indigenous Peoples was already a well-established tradition in settler-colonial warfare long before it became a legal category. And in the summer of 1833, Darwin set out on an overland journey through one of the lesser-known colonial war zones of the Americas, in which the Argentinian army had finally undertaken the *entrada general* (general offensive) against the Indians of the south that their colonial predecessors had rejected.

* * *

The architect of this campaign was General Juan Manuel de Rosas, one of the most controversial figures in Argentinian history. Between 1829 and 1852, Rosas ruled the Argentine Confederation almost without interruption with an iron hand that earned him a reputation as a uniquely monstrous tyrant. Twentieth-century Argentinian historians have been more forgiving. Some have described him as a patriot and nationalist who imposed his authority on a country that was almost ungovernable and defended its interests against an array of foreign enemies that included Britain, France and Brazil. Even the Uruguayan leftist Eduardo Galeano, no friend of dictators or generals, once claimed that 'the black legend that was later spun around his name cannot conceal the national and popular character of many of his administrative measures'.[3]

Rosas was born into a wealthy Buenos Aires cattle-ranching family and grew up in the violent world of the Argentine frontier. His uncle was killed by Indians, and Rosas spent his childhood on his maternal ranch north of the Río Salado, surrounded by gauchos and Indians, who taught him their languages. As a rancher and a politician, Rosas cultivated the common touch. An expert horse-rider, he ate with his peons (unskilled farm workers), recited gaucho poetry and used the knife and the lasso with equal proficiency. He enjoyed good relations with the Afro-Argentines who worked on his estates and made up some 20 per cent of the population of Buenos Aires. Rosas had his peons whipped or stretched out on stakes in the sun for breaking his rules and regulations, and he also had himself whipped for breaking the same laws. He was an astute businessman and ranch manager who acquired large estates in the province of Buenos Aires by the age of thirty, co-partnered a successful salting plant near the Río Salado and formed his own private militia-army known as the Colorados del Monte.

Initially formed to defend his estates against Indian attacks, the Colorados went on to play its part in the warfare between the provincial *caudillos* (strongmen) who vied for power and influence in post-independence Argentina. Rosas was a firm adherent of provincial autonomy against the 'unitarians', who sought to establish the Argentine Confederation as a centralized republic. He was a reactionary authoritarian, hostile to any manifestation of popular sovereignty, who believed that his countrymen were too volatile to form a viable republican government. In 1829, he was appointed general commander of the rural militias of the province of Buenos Aires. Following his victory over the unitarian general Juan Lavalle, he became provincial governor of Buenos Aires from 1829 to 1832. In November that year, the 'Restorer of Laws' presented the Buenos Aires Assembly with a proposal for an offensive against the Indigenous tribes of the southern 'desert' in response to a spate of raids on white settlements by Ranquel and Pampean Indians.

By extending his province's frontiers, Rosas hoped to acquire lands for himself and his province and further his ambitions to become the permanent ruler of the United Provinces. This expedition was hailed as a pivotal moment in Argentine history by his own countrymen. The *British Packet*—a newspaper serving the British mercantile community in Buenos Aires—lamented the occupation of 'a vast and fertile territory ... by a small and scattered race of Indians' and predicted that 'if the present expedition succeeds, and adequate protection be given to settlers, the province of Buenos Aires alone might, at no very distant period, become a State equal to some European kingdoms'.

Rosas prepared for the campaign with characteristic thoroughness. His agents scoured the provinces to buy horses and cattle for his army. He wrote letters to friendly caciques offering rations and other rewards in exchange for their loyalty and cooperation, and he asked the Chilean army to participate in the campaign to prevent his opponents from seeking refuge across the Andes. Though he placed the powerful *caudillo* Juan Facundo Quiroga in overall command of the campaign, there was no doubt in his or anyone else's mind about who was in charge of operations. In March 1833, some 4,000 men marched southwards in three columns from Mendoza in the west to the Atlantic coast. The largest column was commanded by Rosas himself, who led 1,500 men, equipped with artillery, towards Bahía Blanca and the Río Colorado, along the Atlantic coast. Rosas also established a communication system of twenty-one *postas*, equipped with riders and horses, to carry messages back to Buenos Aires and maintain contact with his wife and his political supporters in the city. Land surveyors, priests, engineers, meteorologists and astronomers accompanied an expedition that Rosas promised would 'exterminate the barbarians' and open up 'new channels of commerce' in the 'fertile lands, now populated by savages'.[4]

The expedition quickly ran into the same difficulties that had undermined so many previous attempts. The Chilean contingent was forced to return to Santiago to put down a rebellion. The Division of the Centre was commanded by a general named José Ruiz Huidobro, who was more well known as a comedian and theatre director than for his military prowess. Travelling south from Córdoba with an entourage of actresses, musicians and buffoons,

Huidobro's columns were repeatedly attacked by some 800 Indians under the command of the Ranquel cacique Yanquetruz. When the Ranquels made off with their food supply, Huidobro's soldiers were forced to eat their horses. Others were killed or drowned in quicksand (*guadal*). So many soldiers deserted Huidobro's columns that the government of Córdoba punished anyone who helped deserters with a 100-peso fine and 200 lashes.

Rosas's columns had more success. On 10 May, Rosas reached the Colorado River, where he established his field headquarters. On 9 July, the Argentinian flag flew for the first time from the island of Choele Choel on the Río Negro, after soldiers under the command of General Juan Pacheco waded across the frozen river using their rifle butts to break the ice. That month, Pacheco sent a message to Rosas, warning that an 'English corvette' appeared to be exploring the coast and the river. Neither Pacheco nor Rosas had heard of the *Beagle* or Charles Darwin, but Darwin was very much aware of the general and his campaign, as he prepared to undertake one of his riskier journeys into the South American interior.

* * *

The expedition's motives were personal and scientific. Darwin was keen to make a study of the geology and the flora and fauna of the Pampas, and a journey into the Argentinian interior also offered the prospect of an overland adventure and an escape from seasickness and the cramped conditions on board ship. Darwin had already obtained authorization for the expedition in Buenos Aires in November 1832, before the *Beagle* set out for its first visit to Tierra del Fuego and the Falklands/Malvinas; but he now required a permit from Rosas himself in order to travel further. On 3 August 1833, he was dropped off at the mouth of the Río Negro, 170 miles south of Bahía Blanca. Accompanied by a gaucho escort and an English merchant named James Harris, he rode 18 miles to the fortress/town of Carmen de Patagones—the southernmost outpost of civilization in South America—where he observed 'the ruins of some fine "estancias", which a few years since had been destroyed by the Indians'.

Darwin was also told of local settlers who had fended off 'Araucanians from the South of Chile; several hundreds in number, and highly disciplined'. These raids were becoming rarer, he observed, because 'the government of Buenos Ayres equipped some time since an army under the command of General Rosas for the purpose of exterminating them', and 'the country was thus pretty well cleared of Indians'. On 13 August, Darwin reached Rosas's camp on the Río Colorado. Four days later, Rosas's secretary wrote in his campaign diary that the 'naturalist Señor Carlos Darvien [sic]' had arrived from Carmen de Patagones four days previously, with a passport from the political and military command at Patagones and a note from the minister of war in Buenos Aires, seeking permission to continue to Bahía Blanca, where he was to be reunited with the *Beagle*. The secretary also noted that Rosas had allowed 'el Señor Darwin' to continue his journey with 'all the necessary recommendations to provide him with the assistance he required'.

The only record of the meeting between the gentleman-scientist and the governor-general comes from Darwin's own journal. Darwin spent two days at the camp among Rosas's 'villainous Banditti-like army' in its uniform of sandals, red caps, cloaks and ponchos and the Indian allies who made up a third of Rosas's forces. Darwin heard many stories of Rosas's charisma, horsemanship and the 'despotic powers' he wielded over his soldiers. Such was the loyalty Rosas inspired, Darwin wrote, that when one of Rosas's soldiers murdered a man who 'spoke disrespectfully' about the general, he was released without charge. One of Rosas's Afro-Argentine buffoons described how he had once been staked out in the sun for irritating the general—a punishment the buffoon cheerfully accepted. Darwin described Rosas as

> a man of extraordinary character; he has at present a most predominant influence in this country & probably may end up by being its ruler. He is said to be owner of 74 square leagues of country & has about three hundred thousand cattle. His Estancias are admirably managed, & are far more productive of corn than any others in the country.

On 15 August, Darwin was invited to meet Rosas and found him 'enthusiastic, sensible, and very grave. His gravity is carried to a high pitch.' Darwin left the meeting 'altogether pleased with my interview with the terrible General. He is worth seeing, as being decidedly the most prominent character in S. America.' Darwin would later revise this view, but in the short term, Rosas provided him with a safe conduct, the use of his *posta* system, an offer of accommodation on his ranch and the promise of assistance in transporting fossils back to England. As a result, Darwin was able to travel in relative safety through a colonial war zone that was unlike anything he had ever seen. A contemporary painting of the 'expedition to the deserts of the south' shows 'el gran Rosas' resplendent in his braided cap and jacket on a black horse, surrounded by Indian corpses. Rosas is talking to one of his officers, who points to rows of cavalry charging a horde of fleeing Indians. There were very few, if any, 'battles' of this scale in a campaign that consisted largely of skirmishes and massacres. Rosas's men tortured prisoners to obtain information, and they took few prisoners. In an extant campaign letter, Rosas told one of his officers to take no more than four prisoners at a time. The remainder were to be taken away from the camp and executed 'while trying to escape'.

Rosas's army was barely six months into these operations when Darwin set out with his safe conduct and escort of soldiers, following Rosas's line of *postas* through a landscape of war and violence. On 16 August, he spent the night at a *posta* where a 'Negro Lieutenant born in Africa' and four soldiers had narrowly avoided being massacred some time before their arrival because a band of Indians travelling at night had passed by without seeing them. The following day, Darwin and his escort were forced to leave the road and make their way along a marsh on approaching Bahía Blanca after a 'great gun' from the fortress announced the presence of hostile Indians in the vicinity. While waiting for the *Beagle* to return, Darwin resumed his excavations at Punta Alta, where he was told that 'Indians had murdered every soul' in one of Rosas's outposts. Shortly afterwards, a column of 300 men arrived at the protective fortress with orders to massacre the tribe responsible. On 26 August, Darwin was reunited with the *Beagle* and regaled FitzRoy and the crew with 'travellers' tales' of his adventures.

By this time, Darwin had decided to ride another 500 miles to Buenos Aires. While making his preparations, he heard 'many curious anecdotes respecting the Indians', such as an account of a battle in which 112 Indians had been killed or captured by Rosas's troops, including women and children. 'The Indians are now so terrified that they offer no resistance in body, but each escapes as well as he can, neglecting even his wife & children,' Darwin wrote:

> The soldiers pursue & sabre every man. Like wild animals however they fight to the last instant. One Indian nearly cut off with his teeth the thumb of a soldier, allowing his own eye to be nearly pushed out of the socket. Another who was wounded, pretended death with a knife under his cloak, ready to strike the first who approached.

One of Darwin's informants told him how he had pursued an Indian on his horse and then dismounted to cut his throat with his sabre. All this constituted 'a dark picture,' Darwin wrote,

> but how much more shocking is the unquestionable fact, that all the women who appear above twenty years old, are massacred in cold blood. I ventured to hint, that this appeared rather inhuman. He answered me, 'Why what can be done, they breed so.' Every one here is fully convinced that this is the justest war, because it is against Barbarians.

* * *

The belief that wars between civilized armies and savages required different methods from wars between civilized states was not unique to nineteenth-century Argentina. The British colonial war theorist Charles E. Callwell argued that 'uncivilized races attribute leniency to timidity', and that 'fanatics and savages ... must be thoroughly brought to book and cowed or they will rise again'.[5] This was very much the aim of Rosas's Desert Campaign. But what did Darwin himself think of this 'dark picture'? His responses are not straightforward. 'Who would believe in this age in a Christian, civilized country that such atrocities were committed?' he lamented, in a diary entry in early September. 'The children of the Indians are saved, to be sold or given away as a kind of slave, for so long a time as the owner can deceive them.' The staunch abolitionist nevertheless

insisted that there was 'little to complain of' in such treatment—
an argument he had previously rejected when FitzRoy had made
it in relation to the treatment of slaves in Brazil. He correctly saw
Rosas's war as the beginning of a longer campaign aimed at clearing
the Pampas of its Indigenous inhabitants. 'In another half century I
think there will not be a wild Indian in the Pampas North of the Río
Negro,' he wrote. 'The warfare is too bloody to last; The Christians
killing every Indian, & the Indians doing the same by the Christians.'

Despite his disgust at the cruelty of Rosas's war, Darwin saw
its practical utility. Leaving Rosas's headquarters on 16 August, he
described the rows of willow trees near the camp as useful 'for the
Estancias which General Rosas intends making there' and mused
that 'this war of extermination, although carried on with the most
shocking barbarity, will certainly produce great benefits, it will at
once throw open four or 500 miles in length of fine country for the
produce of cattle'.[6] In another entry on 8 September, he predicted:
'If this warfare is successful, that is if all the Indians are butchered,
a grand extent of country will be in the hands of white Gaucho
savages instead of copper-coloured Indians, the former being a little
superior in civilization, as they are inferior in every moral virtue.'

For all its undoubted brutality, the war was not quite the 'war
of extermination' Darwin believed it to be. Though Rosas had no
qualms about killing Indians who opposed him, he was prepared
to negotiate with caciques who accepted his authority, and he kept
the agreements he made. Darwin became more sympathetic to the
'terrible general' as he drew closer to Buenos Aires. As a guest at
Rosas's ranch at Los Cerrillos on 18 September, he was impressed
by the 'immense herds of cattle' on Rosas's estates, where some
300 to 400 peons have 'defied all the efforts of the Indians'. At the
nearby town of Guardia del Monte, where Rosas had begun his
campaign, Darwin was struck by 'the enthusiasm for Rosas, and for
the success of this "most just of all wars, because against Barbarians".
It is however natural enough, for even here neither man, woman,
horse or cow was safe from the attacks of the Indians.'[7]

Darwin's attitude towards the 'barbarians' was equally
ambivalent. In a long diary entry summing up his travels between 4
and 7 September, he compared Indians who refused to surrender to
'wild animals'. In the same entry, he described an incident in which

three Indian prisoners were shot to death one after another because they refused to divulge the whereabouts of their camp. According to Darwin, the third prisoner told his interrogators to 'fire, I am a *man* & can die'—a gesture of defiance that led Darwin to describe these prisoners as 'noble patriots, not a syllable would they breathe to injure the united cause of their country!'[8] Darwin also wrote of the cacique Chocorí, who escaped from Rosas's soldiers at Choele Choel with his young son by hanging from one side of a saddle-less horse while under fire—an incident that left 'a fine picture' in Darwin's mind of 'the naked bronze like figure of the old Indian with little boy, riding like a Mazeppa on the white horse, thus leaving far behind the host of his pursuers'.[9]

Were these Indians acting on animal instinct? Or were they patriots, fighting and dying to defend their homeland, like the gallant Ukrainian hetman (military leader) from Byron's celebrated narrative poem *Mazeppa*, whom Darwin presented to his readers? Darwin does not seem to have made up his mind. At Rosa's headquarters on 14 August, he saw men of 'a tall exceedingly fine race' among Rosas's Tehuelche allies, in whom 'it is easy to see the same countenance, rendered hideous by the cold, want of food & less civilization, in the Fuegian savage'. Though 'some authors in defining the primary races of man, have separated these two classes of Indians, I cannot think this correct,' he wrote. 'Amongst the young women or Chinas, some deserve to be called even beautiful.'[10] These observations suggest that Darwin did not see savagery as a fixed racial condition but as a cultural state that could be altered through exposure to civilization—a transformation he had already witnessed among the Fuegians.

There was no room for such appreciation in his description of a recently arrived party of Rosas's Indian auxiliaries at Bahía Blanca on 25 August. 'It was impossible to conceive anything more wild and savage than the scene of their bivouac,' he wrote. 'Some drank till they were intoxicated; others swallowed the steaming blood of the cattle slaughtered for their suppers, and then, being sick from drunkenness, they cast it up again, and were besmeared with filth and gore.'[11] Darwin completed this lurid picture with a Latin quote from Virgil's *Aeneid*, describing the Cyclops gorging on the blood of his victims. Other races and racial mixtures were equally unappealing, such as the 'old Chilean miner of the color of mahogany, & the other partly a mulatto' at one of Rosas's *postas*, whom he called 'two such mongrels, with such detestable expressions I never saw before'.

* * *

In September, Darwin returned to Buenos Aires on horseback, weighed down with specimens of birds and animals, after what had been both a harrowing and disturbing journey through a war zone and an epic adventure in the Argentinian interior. In a letter to his cousin, William Darwin Fox, he gave thanks to Providence that 'I have returned with an uncut throat; both Indians & those miscalled Christians, have most carotid-cutting faculties.'[12] That same month, he set out on another fossil-hunting expedition to the province of Santa Fe, following the line of the Río Paraná, where he saw further evidence of the conflict between settlers and Indians. On 2 October, he saw 'some houses now deserted from having been plundered' by Indians, in addition to 'a spectacle, which my guide looked at with great satisfaction, viz the skeleton with the dried skin hanging to the bones, of an Indian suspended to a tree'. The following day, he arrived in Santa Fe, the fiefdom of the *caudillo* Estanislao López, the provincial governor and Rosas ally whose 'chief occupation is killing Indians: a short time since he slaughtered 48 of them. The children are sold for between 3&4 pound sterling.'

This was Darwin's last experience of Argentina's Indian wars. On 21 October, he returned to Buenos Aires to find the city in the midst of a rebellion against the incumbent governor. Though Darwin did not believe the revolt had been instigated by Rosas himself, he predicted that 'Rosas ultimately must be absolute Dictator ... of this country'—a prediction that proved to be correct. By the autumn of 1834, Rosas had signed treaties with the dominant caciques of the south and annihilated or dispersed those who still opposed him. On 25 March 1834, Rosas gathered his army on the banks of the Napostá Grande stream near Bahía Blanca and formally announced that the campaign was over. The *Gaceta Mercantil* listed 3,200 Indians killed, 1,200 of both sexes taken prisoner and more than 1,500 captives freed. Given Rosas's preference for cutting the throats of prisoners 'in order not to waste powder on vultures', these figures were almost certainly an underestimate. A grateful Assembly nevertheless hailed the 'intrepidity, constancy, and patriotism of the brave men' whose efforts had 'at last realized the hopes of centuries, and shut the door against the depredations of the savages who have desolated our fields, banishing thence the population, security and peace'. On 7 March 1835, Rosas was re-elected governor of Buenos Aires and invested with the dictatorial powers he had asked for, and by that time Darwin was on his way back home.

6

THE EMPIRE AND THE FLAG

Even the most observant travellers see the countries they visit through a prism of ideas, prejudices and expectations that inevitably shape their judgements about what is strange, exotic, familiar or distant, and Darwin was no exception. Darwin was an astute, insightful and curious observer of people and places, but his ethnological and anthropological observations owed more to his own subjective impressions than to any obviously scientific criteria, and these observations were often made in passing. In his journal, he admitted that he spent 'but a short time in each place' and that brevity obliged him to draw 'mere sketches instead of detailed observation'—a context he rightly regarded as a limitation. But brevity is not the only explanation for his frequently reductionist and superficial judgements about the Indigenous Peoples he encountered. In a chapter on the 'great world' of London in *The Criminal Prisons of London and Scenes of Prison Life* (1862), the social reformer Henry Mayhew wrote: 'If Arabia has its nomadic tribes, the British Metropolis has its vagrant hordes as well. If the Carib Islands have their savages, the English Capital has types almost as brutal and uncivilized as they. If India has its Thugs, London has its garotte men.'[1] The publisher's introduction praised Mayhew for shedding light on the 'dangerous classes' that consisted of 'men and women who were little higher than Hottentots in the scale of civilization'.[2]

These were not the men and women Darwin had any familiarity with. His ethnological 'sketches' were the work of a highly intelligent and intellectually curious upper-middle-class Englishman with liberal inclinations who had spent most of his life among people of his own class and who travelled the world at a time when Britain was just beginning to acquire its 'liberal empire'. In a parliamentary speech in favour of the renewal of the East India Company's Charter in 1833, the MP, historian and legal member of the Council of India Thomas Babington Macaulay exhorted the British government to educate Indians in the English language as a moral and philanthropic duty, which 'as a people blessed with far more than an ordinary measure of political liberty of intellectual light, we owe to a race debased by three thousand years of despotism and priestcraft'. Macaulay depicted the East India Company as the instrument of a wider civilizing project based on 'the pacific triumphs of reason over barbarism; that empire is the imperishable empire of our arts and our morals, our literature and our laws'.[3]

Darwin shared these convictions, and they were strengthened by his travels. But this did not mean he was blind to the faults of his countrymen. Observing the misery of Brazilian slaves in Bahia following his arrival in South America in March 1832, he railed against the 'polished savages in England' who saw slaves as 'hardly their brethren, even in God's eyes'. But he saw the 'barbarians' and 'savages' of South America, the South Seas and Australia through the eyes of an Englishman, for whom his own country stood at the apex of civilizational achievement in arts, morals, science, literature and law, and he also measured the cultural level of non-English peoples in terms of their proximity to these achievements.

* * *

This sense of cultural superiority permeates Darwin's letters and journal. Returning from Rosas's war in September 1833, he wrote to his sister Caroline from Buenos Aires to describe his eventful journey 'through a district, till very lately never penetrated except by the Indians & never by an Englishmen'. Darwin was then staying at the English merchant Edward Lumb's house, and he told his sister how strange it felt to be 'writing in an English furnished room, & still more strange to see a lady making tea' after the 'wildness & novelty' of his journey through the war-torn Pampas.[4] In another letter to Caroline from Valparaíso in August 1834, he praised an old school friend-turned-shipping merchant named Richard Henry Corfield, who had put him up at his house, and expressed his relief on meeting 'such a straitforward [sic]—thorough Englishman, as Corfield is, in these vile countries'.[5]

Darwin was often hosted by Englishmen during his South American travels, and these English homes were outposts of civility and civilization, where he could drink tea and escape from the 'vile countries' of the New World, whose inhabitants he often described in disparaging terms. In July 1832, he expressed his disgust at the corrupt 'slave society' he encountered in Brazil, whose population 'possess but a small share of those qualities which give dignity to mankind. Ignorant, cowardly & indolent in the extreme.' In Argentina, he described the corruption of local officials, the absence of basic standards of justice and the cruelty and violence of the 'semi-civilized' gauchos. Sailing down the Río Paraná on 16 October 1833 during his fossil-hunting expedition to Santa Fe, he lamented the scarcity of other vessels on the river, which he attributed to the Paraguayan dictator, Dr Francia, and the legacy of Spanish colonial rule. 'How different would have been the aspect of this river if English colonists had by good fortune first sailed up the Plata!' he exclaimed in his journal. 'What noble towns would have occupied its shores!'[6] Returning to revolution-torn Buenos Aires to find his English host's ranch ransacked, he wrote to his sister Caroline: 'Oh these Creoles are such a detestably mean unprincipled set of men, as I hope this world does not contain the like. There literally is only one Gentleman in Buenos Aires, the English Minister.'[7]

In May 1832, in Rio de Janeiro, he attended a religious service on board the British battleship *Warspite*. Watching its 650-man crew take off their hats to sing God Save the King, he described the 'feeling of exultation which is not felt at home' on seeing 'when amongst foreigners, the strength & power of one's own Nation'. Darwin was impressed by the *Warspite*'s technology, its seamless organization, the intellectual curiosity of its officers, the quality and range of music played by the ship's band and the 'cleanliness & extreme neatness' of its storerooms. In effect, the *Warspite* was a microcosm of England itself: a clean, cultured and well-organized and well-governed country that constituted the benchmark for his assessments of the countries he visited. Between these heights of Englishness and the lowliness of the Fuegians, Darwin measured peoples and countries on a sliding scale of savagery and civilization.

Though the South Sea Islanders were 'comparatively civilized', Southern African tribes 'prowling about in search of roots, and living concealed on the wild and arid plains, are sufficiently wretched'. A group of 'gigantic Patagonians'—presumably Tehuelche—at Cabo Gregorio 'behaved like perfect gentlemen, helping themselves with knives, forks, and spoons' and had 'so much communication with sealers and whalers, that most of the men can speak a little English and Spanish; and they are half-civilised, and proportionally demoralised'. Of the natives of Chiloé Island off the west coast of Chile, Darwin observed: 'It is a pleasant thing to see the aborigenes [*sic*] advanced to the same degree of civilization, however low that may be, which their white conquerors have attained.'

These 'white conquerors' were still some degrees below the English, and this patriotism—or chauvinism—becomes even more pronounced in the last part of the *Beagle*'s voyage, when Darwin visits British colonies or countries subject to British influence. 'All the fragments of the civilized world, which we have visited in the southern hemisphere, all appear to be flourishing; little embryo Englands are hatching in all parts,' he wrote of the Cape of Good Hope in June 1836, while Sydney was a 'most magnificent testimony to the power of the British nation: here, in a less promising country, scores of years have effected many times more than centuries in South America. My first feeling was to congratulate myself that I was born an Englishman.' In New Zealand, Darwin compared the neat, whitewashed houses of English colonists and their flowered gardens to native 'hovels' that were 'so diminutive & paltry that they can scarcely be perceived from any distance'. At the missionary settlement of Waimate, he contrasted the 'clean tidy & healthy appearance' of the Maori women who worked as maids to the 'women of the filthy hovels' he had seen elsewhere. Cleanliness and order were always essential hallmarks of civilization—and once again Darwin rarely encountered anything else in his own country. Darwin saw English missionaries as the key to this transformation, whose attention to the 'arts of civilization' had brought 'cordiality' and 'happiness' to a society characterized by 'cannibalism, murder & all atrocious crimes'.

* * *

For both the agnostic Darwin and the devout FitzRoy, missionaries were indispensable instruments of British/English 'soft' power. In September 1836, the two of them wrote a joint letter to the *South African Christian Recorder* denouncing the negative depictions of missionaries at the Cape of Good Hope and in earlier accounts of the South Seas. In *A New Voyage round the World* (1830), the Russian explorer Otto von Kotzebue accused missionaries of infantilizing Tahitians and practising 'a religion which forbids every innocent pleasure and cramps or annihilates every mental power'.[8] Darwin and FitzRoy rejected these criticisms and cited the transformation of FitzRoy's captive Fuegians from 'cannibal wretches' into 'well behaved, civilized people, who were very much liked by their English friends' as evidence that 'a savage is not irreclaimable, until advanced in life'. They praised the English missionaries who had arrived in Tahiti in the early nineteenth century for having induced temperance among the native population and for ending the practices of human sacrifice and infanticide, in 'bloody wars where the conquerors spared neither women or children', and for religious services whose 'appearance was quite equal to that in a country Church in England'—the highest praise.

Unlike the 'wild' Fuegians, Darwin found in the Tahitians 'a mildness in their faces, which at once banishes the idea of a savage, & an intelligence which shows they are advancing in civilization'.[9] In New Zealand, the 'expressions' of the Maori compared poorly with the Tahitians, bringing 'conviction to the mind that one is a savage, the other a civilized man'. Yet even the most wretched Maori compared favourably with the Fuegians ('If the state in which the Fuegians live should be fixed as zero in the scale of governments, I am afraid the New Zealand would rank but a few degrees higher'). In New South Wales, Darwin was struck by the sight of a group of partly clothed Australian Aborigines, some of whom spoke enough English to suggest they might be 'far from such utterly degraded beings as usually represented'—a development that placed these 'harmless savages ... some few degrees higher in civilization, or more correctly a few lower in barbarism, than the Fuegians'.

As was the case in South America, such conclusions were often based on visual impressions. Unable to speak the same language as most of the Indigenous Peoples he encountered, Darwin often made ethnological judgements based on their 'expressions' and the racial/cultural associations they evoked in him. Invited by the local whites at King George's Sound in Tasmania to observe a 'great dancing party' of Aborigines in March 1836, he described an exotic touristic spectacle in which 'the naked figures viewed by the light of the blazing fires, all moving in hideous harmony, formed a perfect representation of a festival amongst the lowest barbarians'. Darwin also relied on local colonial officials and white settlers in reaching these conclusions, such as his claim that Australian Aborigines had 'rapidly decreased' as a result of the 'drinking of Spirits, the Europaean [sic] diseases ... & the gradual extinction of the wild animals'. This decline, Darwin suggested, was primarily due to Aboriginal cultural practices such as infanticide, which supposedly preserved their food supply during their 'wandering life'—a phenomenon he compared to 'what takes place in civilized life, where the father may add to his labor without destroying his offspring'.

Darwin had no evidence that Aborigines killed their children to survive beyond what he was told by white settlers, who had their own interests in making such claims. In August 1824, the colonial governor declared martial law in the Bathurst region of Tasmania in response to a series of violent confrontations between white settlers and local Aborigines. This conflict quickly descended into a vicious settler-colonial war known as the 'Black War' that was not entirely different from the 'war of extermination' that Darwin would later witness in the Pampas. While Aborigines carried out guerrilla raids on isolated settlers who occupied their lands, settler militias, soldiers and convicts collected bounties for the 'black crows' they killed or captured with the specific aim of 'extirpating' and 'exterminating' the Indigenous population.

In 1828, the colonial government introduced martial law across the island of Tasmania, and in 1830 the authorities organized a human dragnet akin to a game hunt, in which a line of soldiers and convict and settler auxiliaries swept through Tasmania in an attempt to eliminate the last pockets of resistance. When these efforts failed, the authorities adopted a policy of wholesale deportation, which effectively eradicated Aboriginal Tasmania. As Darwin observed during a visit to Hobart in February 1836: 'The Aboriginal blacks are all removed & kept (in reality as prisoners) in a Promontory, the neck of which is guarded. I believe it was not possible to avoid this cruel step; although without doubt the misconduct of the Whites first led to the Necessity.' In Cape Town, Darwin was sympathetic to the 'ill-treated aboriginals of the country', whom he regarded as victims of white settlers. But in Tasmania, he accepted the destruction of the Aboriginal population as a 'necessity'. Whatever his misgivings about the initial 'misconduct of the Whites', he celebrated Australia as a 'rising infant [that] doubtless some day will reign a great princess in the south'—a development he attributed to the 'philanthropic spirit of the English nation'.

Faced with the 'march of improvement' in the South Seas, and the imminent transformation of Australia into a 'grand centre of civilization', Darwin concluded his journal with the triumphant affirmation: 'It is impossible for an Englishman to behold these distant colonies, without a high pride and satisfaction. To hoist the British flag seems to draw as a certain consequence wealth, prosperity and civilization.' Many of Darwin's readers would have been pleased to hear such sentiments. On 4 October 1836, the *Beagle* anchored in Plymouth, and the following day Darwin slipped into his family home in Shrewsbury, while his father and sisters were eating breakfast, to the domestic world he had longed for, and sometimes experienced, in the houses of Englishmen during his five years abroad. And no sooner had the traveller returned to his native land than he began to explain the places and peoples he had seen to his contemporaries.

* * *

These explanations and observations catered to an English-speaking readership that shared Darwin's assumptions of British cultural superiority. In the summer of 1839, the extended version of Darwin's travel diaries was published as the third and final volume of the *Narrative of the Surveying Voyages of the Adventure and the Beagle*. The first volume was written by Lieutenant Phillip Parker King, the overall commander of the *Beagle*'s first voyage, and included contributions from FitzRoy and Pringle Stokes. The second was written by FitzRoy and described the voyage Darwin had accompanied. Darwin's own contribution to the trilogy was entitled *Journal and Remarks, 1832–1836*. Mortified by FitzRoy's attempts to introduce his own 'biblical' geology into this collective narrative, Darwin went on to publish his own account separately under the title *Journal of Researches into the Geology and Natural History of the Various Countries Visited by H.M.S Beagle* that same year. This is the book that first established Darwin's reputation, and it remains a classic of the scientific/adventure genre. Bolstered by the approbation of Darwin's hero Humboldt, and by positive reviews from key journals and scientists, the *Journal and Remarks* acquired an international popularity that it has never really lost.

It was through these three volumes that Patagonia first began to capture the imagination of the English-speaking public, and this was largely due to Darwin's evocative and suggestive prose and his descriptions of the Patagonian landscape as an evolutionary necropolis. Parker King, FitzRoy and Darwin all described the Indigenous Peoples of Tierra del Fuego in detail, but Darwin's observations always carried more scientific weight and were instrumental in bringing the 'savages' of Tierra del Fuego to wider national and international attention. In the *New Monthly Chronicle*, Darwin's publisher Henry Colburn praised Darwin's 'full account of ... tribes whose very nature have been, until now, a matter of dispute among travellers, and of a whole people purely aboriginal, and still intact in their native barbarity'. *Littell's Living Age* similarly described the Fuegians as cannibals who 'according to the author, are very little superior in the scale of intelligence to the higher class of brutes', while *The Hampshire Advertiser* quoted extensively from Darwin's descriptions of the 'Civilized Fuegians' and 'Uncivilized Fuegians' to demonstrate that 'all the horrid places of the old world ... fall short, in dreariness and misery, of the reality found in the new'.

In 1845, Darwin published a second edition of the journal, which included his extended descriptions of the Fuegians based on the observations in his original diary and drew on his 1838 encounter with an orangutan at London Zoo called Jenny or Lady Jane. This was the first time Darwin had actually seen an ape, and it explains his reference to York Minster's skin colour in the second edition, in which the *Beagle*'s crew members 'expressed the liveliest surprise and admiration at its whiteness, just in the same way in which I have seen the ourang-outang do at the Zoological Gardens'.[10] Other readers also drew attention to the connections between the Fuegians and animals. A French food journal described 'the Eskimo, the Fuegians and, more rarely, the Hottentots [who], eat raw meat with an altogether bestial gluttony' as 'singular deviations from the habitual practices of civilization [that] indicate that these nations have fallen to the last degree of mindlessness'.[11] If Darwin's experiences in Patagonia had begun to transform his view of nature, his journal transformed a voyage of scientific investigation into a rousing confirmation of British-led global imperial progress, at a time when the destructive impact of civilization on the world's Indigenous Peoples had become increasingly difficult even for the most fervent imperialist to ignore.

* * *

In the American artist Asher Durand's 1853 painting *Progress (The Advance of Civilization)*, three Native Americans stare down helplessly from a depleted forest towards a lake lined by factories, railway bridges, canals and viaducts, while settlers on loaded wagons make their way past telegraph poles and a railway bridge towards the billowing steeples in the distance. This depiction of Indigenous 'savages' as marginalized outsiders in a country that was being transformed by industrialization and urbanization was very much a reflection of the mid-nineteenth-century world. By 1800, the Native American population of the United States had fallen from an estimated pre-contact population of 1 to 10 million to 600,000. By the 1890s, that figure had fallen to 237,000.[12] Between 1840 and 1896, the Maori population fell from 80,000–90,000 to 42,000. In the same period, the European population of New Zealand increased from 2,000 to 791,00.[13] The same decline occured in other European colonial territories. In 1837, the British abolitionist Thomas Fowell Buxton established a fifteen-strong parliamentary Select Committee to compile evidence on 'the injustices and cruelties with which the Aborigines have hitherto been treated, and the pernicious effects which have resulted to them from their intercourse with European nations'. The committee consulted soldiers, politicians, settlers, missionaries and natives from South Africa, Australia, Canada, New Zealand, British Guiana and the South Seas before concluding, in 1839, that white colonial settlement had been a disaster for

> uncivilized nations ... Too often, their territory has been usurped; their property seized; their numbers diminished; their character debased ... European vices and diseases have been introduced amongst them, and they have been familiarized with the use of our most potent instruments for the violent destruction of human life, viz. brandy and gunpowder.[14]

Among many examples of this 'devastation and ruin', the committee cited the complete elimination of the Indigenous population of Newfoundland, the decimation of the Cree Indian population of Canada from 10,000 in 1800 to 200 in 1837, and the decline of the Hottentot or Bushman population of South Africa from a pre-colonial population of 200,000 to 32,000. The committee referred to the 'extermination' and 'oppression' of 'the natives of barbarous countries' as 'a practice which pleads no claim of indulgence; it is an evil of comparatively recent origin ... it had been a burthen on the empire'.

Faced with these 'frightful calamities' and 'the practice of cruelty and injustice towards our uncivilized fellow-men', the committee insisted on 'the absolute necessity of adopting immediate measures for their protection and preservation'. These recommendations resulted in the foundation of a non-governmental charity-cum-campaigning organization called the Aborigines' Protection Society, under the leadership of the ethnologist and physician Thomas Hodgkin, which set out to 'ensure the health and well-being and the sovereign, legal and religious rights of the indigenous peoples while also promoting the civilization of the indigenous people who were subjected under colonial powers'.

Not everyone believed these objectives were attainable—or desirable. In *The Natural History of the Human Species* (1848), the British army colonel, naturalist and artist Charles Hamilton Smith argued that the 'less gifted tribes' were 'predestined to perish beneath the conquering and all-absorbing covetousness of European civilization'.[15] Though Smith admitted that these outcomes placed 'an enormous load of responsibility on the perpetrators', he concluded that 'their fate appears to be sealed in many quarters, and seems, by a pre-ordained law, to be an effect of more mysterious import than human reason can grasp'.

This sense of tragic inevitability was a recurring theme in what scholars have called the 'doomed race theory' or 'extinction discourse' that attached itself to the world's 'savage' peoples in the nineteenth century. In his account of the third and final *Beagle* voyage to Australia in 1837, Darwin's former shipmate John Lort Stokes lamented the disappearance of Indigenous Peoples in Australasia, whose 'destiny is accomplished'. Though Stokes acknowledged that white civilization bore some responsibility for this outcome, his only proposed solution was a kind of racial palliative care in which 'all we can do is to soothe their declining years, to provide that they shall advance gently, surrounded by all the comforts of civilization, and by all the consolations of religion, to their inevitable doom; and to draw a great lesson from their melancholy history'. In the *Journal of Researches*, Darwin also identified a 'mysterious agency generally at work':

> Wherever the European has trod, death seems to pursue the aboriginal. We may look to the wide extent of the Americas, Polynesia, the Cape of Good Hope, and Australia, and we find the same result. Nor is it the white man alone that thus acts the destroyer; the Polynesian of Malay extraction has in parts of the East Indian archipelago, thus driven before him the dark-coloured native. The varieties of man seem to act on each other in the same as different species of animals—the stronger always extirpating the weaker.[16]

In Darwin's original diary, there is no mention of this 'mysterious agency'. Yet three years after his return to England, he was suggesting that the 'varieties of man' might interact like animals did, and that the 'extirpation' of the weaker variant might be the result of a natural law. In 1839, Darwin was one of the three secretaries from the geological section of the British Association for the Advancement of Science (BAAS) who attended a lecture in Birmingham by the anthropologist James Cowles Prichard on the subject of 'the Extinction of Human Races'. Like the parliamentary

85

select committee, Prichard depicted European colonization as the 'harbinger of extermination of the native tribes' and called for action to prevent the complete destruction of all 'aboriginal nations of most parts of the world' by the end of the century. Following this talk, the BAAS agreed to put aside 5 pounds to print 'a set of queries to be addressed to those who may travel or reside in parts of the globe inhabited by the threatened races'.[17]

This questionnaire covered a range of issues of interest to the scientific community. Travellers were asked to compile details about the height and weight of the peoples they encountered, their complexions, skin colour and hair texture. At a time when 'skull science' was at the centre of scientific discussions regarding racial typologies, the size and shape of Indigenous skulls received particular attention. Profiles and front views were to be measured in accordance with 'the division and terms introduced by craniologists' so as to determine 'the corresponding development of moral and intellectual character'. Prospective researchers were asked to ascertain whether 'a district obviously possesses two or more varieties of the human race' and 'state the result of their intermixture on physical and moral character'.

Darwin wrote some of the questions to this questionnaire, some of which were taken from his notebooks. Even at this early stage in his post-*Beagle* career, Darwin had begun to consider the wider implications for humanity of what he later called 'the quiet war of organic beings, going on in the peaceful woods'. These reflections owed much to his reading of Thomas Malthus (1766–1834) in September 1838. In his *An Essay on the Principle of Human Population*, Malthus argued that human population was subject to natural checks that prevented it from growing beyond its capacity to support itself. Looking back to the disappearance of 'barbarian tribes' in ancient history, Malthus speculated that '[i]n these savage contests many tribes must have been utterly exterminated ... The prodigious waste of human life occasioned by this perpetual struggle for room and food was more than supplied by the mighty power of population, acting, in some degree, unshackled from the consent habit of emigration.'[18]

Many years later, Darwin recalled how his post-*Beagle* reading of Malthus helped him to develop his theory of natural selection with regard to plants and animals and enabled him to frame his understanding of the circumstances under which 'favourable variations [of species] would tend to be preserved & unfavourable ones to be destroyed. The result of this would be the formation of new species.'[19] Darwin was also aware of the implications of the 'struggle for existence' for humanity, in an age in which history was increasingly perceived in precisely these terms. In the first two decades after his return to England, Darwin was more preoccupied with tracing the differences between species than he was with different categories of humanity, but even as he studied barnacles, flowers and pigeon beaks, his mind continued to return to the questions he had first asked himself in Tierra del Fuego. 'I will not shirk from difficulty,' he wrote, in a note on the physical differences between humans and animals and between different animal species in his 1839 Notebook E: 'I have felt some difficulty in conceiving how inhabitant of Tierra del Fuego is to be converted into civilized man—ask the missionaries about Australians yet slow progress has done so. Show a savage a dog, & ask him, how wolf was so changed.'[20] 'A profound consideration of method by which races of men have been exterminated ... is very important,' he wrote in the recovered pages of his 1839–41 'Torn Apart' notebook, citing Pritchard's paper on Aboriginal peoples as a source.[21]

Even in the menagerie of mastodons, dogs, wolves, dogs, wild pigs and apes, the 'savage' continues to appear in his mind as a zoological type and a point of reference in his investigations and reflections on the differences between species and life forms and the disappearance of species. It would be more than three decades before Darwin was ready to comment publicly on these differences, but even as he continued to reflect on them in private, the world was coming closer to the 'uttermost end of the earth', and Patagonia was drawing closer to the world.

PART II

CIVILIZATION AND BARBARISM

Patagonia: what of its climate?
It is cold and inhospitable.
Who are the inhabitants?
Wandering tribes of Indians
To what country do the islands on the Pacific coast
of Patagonia belong?
To Chile.
What is the surface of Patagonia?
It is a barren and stony waste.

Colton and Fitch's Introductory
School of Geography, 1863

7

THE CALIGULA OF THE RIVER PLATE

By the time Darwin's first account of his travels was published, the United Provinces of the Río de la Plata had been reorganized into the Confederación Argentina (Argentine Confederation) under the leadership of the all-powerful governor of Buenos Aires Juan Manuel de Rosas. Having obtained the 'sum of public power' through his Desert Campaign, Rosas governed his country with the same implacable ferocity Darwin had observed at his field headquarters. His instinctive authoritarianism was intensified by a constant stream of political crises that included British and French naval blockades, attempted coups and rebellions and the 'Guerra Grande' (Big War) in the Banda Oriental (Uruguay), where Rosas imposed a nine-year siege on Montevideo in support of his favoured faction. The main threat to Rosas's rule came from the conflict between the 'unitarian' proponents of a single unified Argentinian republic with Buenos Aires as its capital and the 'federalist' *caudillos* who resented the domination of Buenos Aires and favoured provincial self-rule. Though Rosas was from Buenos Aires and a staunch defender of his province, he was suspicious of the liberal inclinations of the unitarians and did not believe that his countrymen were politically capable of establishing a republic.

Instead, Rosas opted for a confederation based on provincial autonomy, with Buenos Aires in charge of customs and foreign policy, and he defended this model with levels of terror and repression that Argentinians would not experience again until the military dictatorship of the 1970s. Thousands of real or suspected political opponents were killed, imprisoned or exiled. Men and women of all ages were flogged, arrested or executed for offences that included not wearing the federalist colour red in public, wearing the unitarian blue or even wearing u-shaped beards that supposedly indicated unitarian sympathies. These were years in which the slogan 'Death to the Savage Unitarians!' was ritualistically proclaimed at every public ceremony and called out by the nightwatchman every hour; in which Rosas's steely eyed portrait stared down from public buildings and church altars or displayed on carts and pulled through the streets by magistrates and other leading citizens in annual festivals that paid homage to the dictator.

The repression reached a peak between 1839 and 1842, when Rosas faced a series of rebellions and an invasion from the 'League of the North' under his old enemy, the unitarian general Juan Lavalle. In response, Rosas unleashed his paramilitary police force, the Mazorca, on the defenceless population. Armed with knives, whips and lances, and dressed in the red caps and ponchos of Rosas's rural militia, Mazorca bands roamed the streets of Buenos Aires, carrying lists of unitarians. With no one but Rosas to restrain them, the Mazorca raped, looted and killed with impunity. Real or suspected unitarians were hanged or had their throats cut on the spot. Afterwards, their heads were paraded in carts, adorned with blue ribbons and their corpses labelled *carne con cuero* (beef with hide). The numbers of deaths may have been exaggerated by Rosas's exiled opponents for propaganda purposes, but the cruelty was not invented. The most famous contemporary painting of Rosas shows him in his brigadier general's uniform, in epaulettes, high braided collar and the red federalist sash. With his imperious gaze and piercing blue eyes, aquiline nose and harsh thin mouth, he looks very much like the 'Caligula of the River Plate' that his enemies depicted.

According to one of his secretaries, Rosas was so fearful of assassination that he posted an Indian as a bodyguard and kept a horse permanently saddled outside his office, where he spent the day 'spurred, whip in hand, with hat and poncho, always ready to mount his horse'. Others found him warm and affable but prone to cruel practical jokes, whether lassoing his friends from their horses or leaving a dead viper near a sleeping visitor before waking him up to tell him he had been bitten. The American *chargé d'affaires* described him as a man of 'magnanimity and moderation' whose popularity among the gauchos recalled American farmers, and whose manner 'has something of the reflection and reserve of our Indian chiefs'.[1] Rosas's common touch was evident at his palatial residence in the northern Buenos Aires district of Palermo, where visitors from all classes and races came and went through grounds populated with ostriches, lamas and exotic birds to petition Rosas or his daughter Manuela, who became his secretary and main confidante following the death of his wife Encarnación in 1838.

One scandalized upper-class *porteño* (resident of Buenos Aires) described a 'slum committee' made up of 'negroes and mulattos, gauchoes and common people, steel shop thugs, [who] went in and came out mixed with soldiers and well-heeled gentlemen'.[2] These visitors also included the Indian caciques with whom he had made treaties during his Desert Campaign. On one occasion, according to the British consul Woodbine Parish, Rosas received a 'large party of some of the friendly tribes, with their wives and children' at his home in Buenos Aires. When some of these visitors showed symptoms of smallpox, Rosas persuaded their companions to accept smallpox vaccinations by showing them the vaccination mark on his arm. It was a measure of their loyalty to Rosas that 150 Indians agreed to be vaccinated, and this trust was largely reciprocated. Beset by political crises and external threats, the Argentine Confederation was in no position to extend the new 'internal frontiers' in the southern 'desert' that Rosas's army had conquered in 1833–4. For much of his turbulent seventeen-year rule, these frontiers remained relatively stable, and this was largely due to the system of alliances Rosas established with a new generation of caciques.

* * *

Within a few years of the Desert Campaign, the army had withdrawn its garrison from the island of Choele Choel, which reverted to its previous role as an Indigenous transportation hub and round-up point for cattle, horses and captives on the 'Chilean road' across the Andes. To the west, in the present-day Patagonian province of Neuquén, the Tehuelche-Mapuche cacique Valentín Sayhueque established an Indian confederacy in the Country of Wild Apples. To the north-east of Neuquén, the Ranquels or 'people of the reeds-grass' dominated the dry Pampas from their 'capital' in Leubucó. The Ranquels were not a homogenous group, and their loyalty to the Argentine state often depended on their individual chiefs. Ranquels had fought both with and against Rosas during the Desert Campaign. In 1834, a group of Ranquels loyal to Rosas attacked the cacique Painé Guor, the successor to Rosas's former enemy Yanquetruz, and captured Painé's nine-year-old son Panguitruz Guor ('Hunter of Lions'). The Ranquels handed their prisoner to Rosas, who adopted the boy as his godson and gave him the name Mariano Rosas. For five years, the Hunter of Lions lived on Rosas's ranches under his given name before escaping one night to his father's *tolderías* in Leubucó. Rosas sent an affectionate letter to his 'beloved godson' inviting him to 'visit me with some friends'.

Panguitruz Guor/Mariano Rosas never accepted his godfather's offer, and though he went on to become one of the most important Ranquel chiefs, in war and peace, he did not attack his godfather. Despite these occasional conflicts, Rosas continued to rely primarily on *negocio pacífico* (peaceful negotiation) with Indigenous caciques to manage the situation in the southern frontiers. In return for refraining from raids and providing military service when required, loyal caciques received monthly deliveries of mares, colts, tobacco, sugar, salt, yerba maté and other commodities. Rosas's rigorous application of these treaties explains the great affection in which he was held by Pampean caciques such as Cachul, who addressed a parade at the garrison-fortress of Tapalqué to celebrate Rosas's third governorship with the words: 'Juan Manuel is my friend. He has never deceived me. I and all my Indians will die for him.'

Rosas also entered into a strategic alliance with the Mapuche chieftain Juan Calfucurá, the founder of the Curá (stone) dynasty and the dominant cacique of the Argentinian south for more than half a century. Born in Chile into an important Mapuche family around 1790, Calfucurá crossed the Andes in 1830 and 1831 and quickly established his authority over the Salinas Grandes region through a combination of violence, diplomacy, strategic marriages, charisma and oratory and a reputation for supernatural powers that he did much to cultivate. Calfucurá later claimed that Rosas had personally invited him to cross the Andes in order to impose order on the south. Whatever the truth, the Mapuche *toki* (war chief) became the pillar of Rosas's system of alliances. In September 1834, Calfucurá's warriors massacred some 200 rival Boroano Indians from Chile who had established themselves in the Salinas Grandes salt flats. From his base in Carhué, near the thermal waters of Lake Epecuén, Calfucurá established an Indigenous 'confederation' of tribes and chieftainships that extended from the Andes to Bahía Blanca on the Atlantic, with Salinas Grandes as its strategic centre.

Calfucurá was intensely suspicious of the intentions of the *huinca* (white man) towards Wallmapu—the historic lands of the Mapuche, which extended from Chile to the Atlantic—but he trusted Rosas completely and zealously defended his interests in the Pampas in exchange for monthly *raciones* (rations) and other gifts, which Calfucurá regularly discussed with the governor-general through his Spanish-speaking *lenguaraces* (secretary-translators). The strategic alliance between Rosas and Calfucurá may also have resulted in the brutal events that followed the transportation of a large group of Ranquel and Boroano prisoners from the Río Negro to Buenos Aires in 1836. According to the anonymous 'British gentleman' who wrote to the British foreign secretary Lord Aberdeen from Montevideo in 1844, these prisoners included 'fathers and sons of eight and nine years old, with their aged grandsires, to the number of nearly one hundred'. On 8 July 1836, all these prisoners were taken from their 'horrible prison in groups of ten and twelve' and 'shot down by their executioners, who fired upon them in platoons without particular aim'. The bodies were then loaded on to carts and taken to a mass grave, where some of the survivors tried to crawl away, only to be 'dragged back, with their throats cut and their brains beat out, by their merciless murderers'.[3]

For the most part, such methods were not required. In 1845, the Scottish merchant William MacCann visited Rosas at his 'pleasure grounds' in Palermo and described his 'handsome, ruddy countenance, and portly aspect' that 'gave him the appearance of an English country gentleman'.[4] MacCann was given permission to travel into the Argentine interior because Rosas was keen for the Scotsman 'to judge for myself of the state of the country ... that the real truth should be told'. MacCann was curious to see 'man in his native condition, free from the deformities as well as the elegances of civilized life'. Though he set out on his journey in the belief that 'the more barbarous races … when the habits of civilized life were presented to them … could not fail to admit their superiority, and speedily to adopt them', these feelings underwent a 'radical change' as he travelled through the frontier zone.

At the garrison-town of Azul, he was disgusted by Rosas's Indian allies, who lived 'in a state of abomination difficult scarcely possible to describe or conceive' among the 'carcasses of horses … lying about in various states of putrefaction'. MacCann struggled to believe that 'the different races of man' were descended from the 'same parent stock'. He noted with satisfaction that the 'greater part' of Argentina's Indigenous Peoples 'have become extinct, and the remainder are rapidly passing away', and he saw these developments as a harbinger of the 'rapid extinction of the feeble races of the new world' when 'the hundreds of races, with their millions of people, that lived in the hemisphere … shall have disappeared, and their name and their language will be forgotten'.

* * *

The Rosas regime made only sporadic attempts to assert Argentina's interests in Patagonia and Tierra del Fuego in this period. In 1832, Rosas appointed a governor to establish a penal settlement in the Falklands/Malvinas in response to British attempts to colonize the islands. Rosas's representative was murdered in a mutiny shortly after his arrival, and the following year the British reoccupied the islands with the assistance of FitzRoy and the *Beagle*. In 1849, Rosas signed the Convention of Settlement with Britain, which effectively resolved the dispute over the Falklands/Malvinas in return for British recognition of Argentinian sovereignty in the River Plate and the Río Paraná. As a result, the fortress/settlement at Carmen de Patagones continued to constitute the sole outpost of Argentinian national sovereignty on the northern boundary of Patagonia at the mouth of the Río Negro. But Rosas's political enemies had begun to imagine a very different Argentina that would follow the end of his regime, in which all the country's 'deserts' would be conquered and civilized.

To the exiled writers and intellectuals known as the 'Generation of '37', associated with the Buenos Aires literary salon organized by the writer and bookshop owner Marcos Sastre in 1835, the main obstacle to achieving these aims was Rosas himself. Politically liberal and steeped in the French philosophical school known as positivism, with its emphasis on scientific knowledge, reason and the teaching of science, these writers sought to build a Europeanized Argentina purged of the 'gauchoesque' traditions epitomized by Rosas and provincial quasi-military strongmen. Their writings are sprinkled with disdainful references to the 'rural chieftains', 'poncho-clad barbarians' and the gauchos and poor blacks whom Rosas cultivated so assiduously. In Esteban Echeverría's short story *El matadero* (The slaughter-yard, 1871), a haughty unitarian gentleman is taken for a 'gringo' by the *rosista* workers in the Buenos Aires slaughterhouse, who drag him from his horse and submit him to a grotesque mock-trial before cutting his throat. The story was probably written in 1837, and its message was clear: the Rosas regime was the expression of the most primitive, barbaric and xenophobic instincts of the lowest of his countrymen, steeped in blood, offal and gratuitous slaughter.[5]

There is some irony that the man who had once waged war on Indigenous 'barbarians' in the name of civilization, and who regularly denounced his 'savage' unitarian enemies, was himself depicted as the epitome of barbarism and savagery by his opponents. These accusations found their most eloquent and sophisticated expression in the writings of the exiled journalist and politician Domingo Faustino Sarmiento (1811–88), one of the towering figures of nineteenth-century Argentinian history. In his seminal *Facundo: Or, Civilization and Barbarism* (1845)—a fusion of essay, novel, cultural analysis and political manifesto—Sarmiento depicted Argentina's post-independence history as a struggle between 'two diverse civilizations: one Spanish, European, cultured, and the other barbarous, American, almost indigenous'.[6] Where the former was associated primarily with the city of Buenos Aires, 'barbarous' Argentina was symbolized by the Argentine interior beyond the city, where 'the desert wilderness surrounds it on all sides and insinuates into its bowels' and 'savages lurk, waiting for moonlit nights to descend, like a pack of hyenas, on the herds that graze the countryside, and on defenseless settlements'.[7]

In Sarmiento's histrionic prose, this 'desert' was a stagnant vacuum, waiting to be filled by the 'vivifying fluids of a nation' and 'a million industrious Europeans', who would transform the country so that 'in twenty years, what has happened in North America in the same amount of time will happen here: cities, provinces, and states have been raised in deserts where, not long before, herds of wild bison grazed'. These objectives required both the overthrow of Rosas and the elimination of the marauding 'savages, avid for blood and plunder' who opposed the settlement of Argentina's barbarous 'desert'. In a footnote on the 1833–4 Desert Campaign, Sarmiento praised Rosas for 'driving off the untamed barbarians and subjugating many tribes', but he also claimed that Rosas's 'inexplicable political system' had halted expeditions further south that might have succeeded in 'evicting the savages' completely.

Sarmiento's attitude to Argentina's Indigenous Peoples was unambiguous. Writing for the Chilean newspaper *El Progreso* in 1845, Sarmiento called the Mapuche territories of Araucanía 'a nation foreign to Chile' that Chile must 'absorb, destroy, enslave, no more or less than the Spanish would have done'. 'By occupying these territories,' Sarmiento argued, Chile's rulers would be 'simply doing what every civilised people does with the savages ... it absorbs, destroys, exterminates them'.[8] Though Sarmiento admitted that '[i]t may seem unjust to exterminate savages, to suffocate nascent civilizations, to conquer peoples that are in possession of a privileged land', it was 'thanks to this injustice [that] America, instead of remaining abandoned to the savages, incapable of progress, is today occupied by the Caucasian race, the most perfect, the most intelligent, the most beautiful and the most progressive on earth'. This outcome was a demonstration of the 'immutable laws' of history by which 'the stronger races exterminate the weaker, the civilized peoples supplant the savages in the possession of the land'.

Other members of the Generation of '37 expressed similar sentiments. To the exiled political theorist Juan Bautista Alberdi (1810–84):

Those of us who call ourselves American are nothing more than Europeans born in America; our skull, blood, colour—everything is from [Europe] ... Who among us would not prefer a thousand times over to see his daughter marry an English shoeman rather than an Araucanian prince? In America everything that is not European is barbaric: there is no division other than this one: Indian which is synonymous with savage, and European which means those of us born in America, who speak Spanish and believe in Jesus Christ.[9]

In their language at least, Rosas's opponents were more implacable enemies of Indigenous Argentina than the 'terrible general' himself. In 1849, the *North American Review* described the Argentine Confederation as a collection of 'detached cities with surrounding cultivated tracts, forming, as it were, oases of civilization in the midst of a vast expanse of untilled and almost uninhabited plain'.[10] The article lamented the 'predominance of a provincial over a national spirit' in which 'so strong has been the disposition manifested, by the interior provinces especially, to act upon the most ultra principles of state rights, that the action of the central government has been greatly embarrassed'. From the point of view of Patagonia's Indigenous Peoples, this absence of a 'national spirit' worked to their advantage and limited the central government's ability to extend Argentina's 'oases of civilization' further south.

* * *

Rosas's political trajectory did not go unnoticed by the scientist who had once benefitted from his largesse. In 1845, Darwin's former shipmate, Bartholomew James Sulivan, then the commander of the HMS *Philomel* and of the Anglo-British squadron that sided with Montevideo in its conflict with Buenos Aires, wrote to Darwin of the '[m]ost dreadful butcheries' carried out by Rosas's troops during the siege of the Uruguayan capital, in which hundreds of prisoners were massacred.[11] Reports from Argentinian exiles on Rosas's crimes also found their way into the British press. Such accusations— coupled with the tensions between the British government and the Argentine Confederation—may well have contributed to Darwin's change of heart regarding the 'man of extraordinary character' whom he had once predicted would bring about the 'prosperity and advancement' of his country. The revised second edition of his *Journal of Researches* in 1845 carries the apologetic footnote: 'This prophecy has turned out entirely and miserably wrong.'[12] In his account of Rosas's Desert Campaign, Darwin also noted that '[s]ince leaving the South American continent we have heard that this war of extermination completely failed'.

In 1929, the Argentinian writer and journalist Ricardo Sáenz Hayes found a portrait of Rosas in his military uniform at Darwin's home in Down House bearing the inscription: 'General Rosas, friend and protector of Darwin.' Saenz was surprised to find this portrait of a 'feudal lord' in a 'house dedicated to science', which he attributed to Darwin's 'fetish for the exotic'. The painting still hangs in the former servants' quarters-turned-boardroom of Darwin's house in Kent. It is not known how Darwin acquired it, but the most likely source is Rosas himself. In 1852, the Caligula of the Pampas was finally overthrown by a Brazilian–Uruguayan–Argentine military coalition led by the powerful *caudillo* of Entre Ríos province, Justo José de Urquiza, in which Sarmiento also participated as an officer. Members of Calfucurá's Salinas Grandes Confederacy fought on Rosas's side in the Battle of Caseros on 3 February 1852, but they were not enough to prevent his defeat.

Rosas fled the battlefield, pausing under a tree to write a letter of resignation with his wounded hand. In Buenos Aires, he sought refuge at the house of the British *chargé d'affaires*, who granted him exile in England and helped him escape from Buenos Aires in disguise, accompanied by his daughter Manuela and other members of his family. Rosas and his entourage were taken to Ireland and then to England, where they were greeted in Plymouth on 26 April with an official reception. The *Times* criticized 'the eagerness of English gentlemen, high in military and naval authority, to grasp his blood-stained hand', but despite his conflicts with Britain, the British government was more than willing to accommodate him. Unable to access his vast frozen assets in Argentina, Rosas moved into rented apartments at the Windsor Hotel in Southampton with the assistance of Home Secretary Lord Palmerston, who allowed him to take out a bank loan to pay for them.

When Urquiza temporarily lifted the embargo on his assets, Rosas sold one of his ranches and rented a 5-storey town house in Carlton Crescent. The *Hampshire Advertiser* later recalled how 'the General used to ride through the streets almost daily on a beautiful black horse, and his majestic form and military bearing, together with the trappings of his steed, always attracted much attention and admiration'. Rosas acquired an English maid and a black Argentinian servant, and his house became a regular point of call for a stream of prestigious Latin American visitors passing through Southampton. In 1855, the Chilean journalist and politician Vicente Pérez Rosales arrived in Southampton and asked his innkeeper where Rosas lived, only to be told 'that gallows bird is certainly up to no good; and if he's not killing people here the way he did in Buenos Aires, that's because in England it's just a step from murder to the gallows'.[13] Pérez Rosales spent six days in Rosas's company and later recalled how this 'singular man' was 'rather obsessed with his belief that it was impossible for the Argentinians to live in peace except under an absolute government' and lived in constant expectation of a recall from his countrymen 'with every steamer that reached Southampton, and with every steamer he was disillusioned anew'. These expectations were never realized, as Rosas's opponents returned from exile and began to construct the new state.

8

THE CITY AND THE DESERT

MacCann's travels through the Argentine *tierra adentro* (interior or hinterland) populated by Basque, Irish and English sheep farmers and cattlemen were a demonstration of the relative security of the Argentine frontier zoner in the Rosas years. Beyond the line of forts, ranches and the *tolderías* that formed Calfucurá's Salinas Grandes Confederation, Patagonia remained much as it had been for centuries: a *res nullius* inhabited by 'tribes of wandering Indians' with little immediate appeal to the potential colonist. In 1849, Benjamin Franklin Bourne, the first mate of the American schooner *John Alleyne*, was captured by a party of Tehuelche at Cape Virgenes, at the eastern entrance to the Strait of Magellan. In *The Captive in Patagonia: Or, Life among the Giants* (1853), Bourne described his three months in captivity and eventual escape from the 'monstrous giants' whom he regarded as 'deficient in the morals as in the refinements and courtesies of domestic life; their licentiousness is equal to their cruelty'. According to Bourne, these 'giants' were gambling addicts who used pieces of guanaco skin as cards 'on which are rudely depicted dogs and a variety of other beasts, with divers mystic marks and scrawls'. Bourne believed—inevitably—that his captors were cannibals. He warned potential settlers that

> if the other tribes inhabiting the country resemble that with which I was domesticated, it must be a hazardous enterprise for missionaries to attempt the propagation of the gospel among them ... To live like the savages would be simply impossible to men who have been habituated to the comforts of civilized life.[1]

Bourne's descriptions of the landscape were unlikely to encourage anyone to abandon these comforts. Patagonia was 'bleak, barren, desolate, beyond description or conception—only to be appreciated by being seen'. This was the general view of Patagonia in both Chile and Argentina in the first half of the nineteenth century. An 1849 article in the Chilean newspaper *El Progreso* described the far south as a 'useless thing' and asked what interest Argentina could have in such a 'frigid country, remote and inhospitable'. Despite its lack of any obvious appeal, both the Chilean and Argentinian governments regarded Patagonia as an important national resource in the longer term, and both countries sporadically attempted to assert their rival territorial claims. To Argentina, Patagonia consisted of all the territories west of the Andes and south of the Río Negro, including Tierra del Fuego. Successive Chilean governments argued, on the basis of Spanish colonial documents, that the territories south of the Río Negro on both sides of the Andes belonged to Chile, from the Pacific and the Atlantic oceans.

This dispute lasted for roughly sixty years and at times brought the two countries close to war before it was finally resolved through negotiations in Argentina's favour in the 1881 Boundary Treaty. Neither country recognized any permanent Indigenous jurisdiction over these territories. For this reason, the Chilean historian Alberto Harambour has described nineteenth-century Patagonia as a 'triple borderland' where the interests of Argentina and Chile, white settlers and Indigenous Peoples and European imperial powers all converged.[2] In the first half of the nineteenth century, international interest in the Patagonian 'borderland' was dominated by British commercial and strategic aspirations. In 1835, FitzRoy and possibly Darwin attended a meeting at the home of a British merchant in Valparaíso with the Chilean statesman and entrepreneur Diego Portales and the American entrepreneur William Wheelwright in which Wheelwright discussed the possibility of a postal steamship service in the Pacific operating along the South American coast.

These discussions resulted in the foundation of the Pacific Steam Navigation Company in 1840, which sent two paddle steamers, the *Chile* and the *Peru*, on the first Atlantic–Pacific steamship voyage from Limehouse docks to Valparaíso via the Strait of Magellan that same year. With the advent of steam navigation, the Patagonian south acquired a new international significance. In 1837, Admiral Jules Dumont d'Urville led a naval expedition to Antarctica, which passed through the Strait of Magellan. D'Urville urged France to establish a settlement at Port Famine to pre-empt the British, and in 1840 the French commercial agent in Tahiti drew up a 'Project for the Colonization of Patagonia' with the same objective. In 1843, Clément Adrien Vincendon Dumoulin, a hydrographic engineer on d'Urville's expedition, urged France to establish a permanent base in the Strait of Magellan and suggested that Patagonia had the potential to become the granary of Oceania and South America. These proposals were debated in the French Chamber of Deputies and eventually rejected on grounds of cost and difficulty.

Chilean and Argentinian claims rarely featured in these debates, and the sovereignty of the Indigenous inhabitants of these territories was not recognized by anybody. For both Britain and France, sovereignty over the 'empty' lands of the Patagonian south was still decided by whichever country could establish a permanent physical presence in the region, and 'wandering Indians' and 'savages' did not count. In *The Law of Nations* (1758), the Swiss jurist Emmerich de Vattel argued, specifically in reference to the New World, that it was lawful for any state to take possession of a 'vast country, in which there are none but erratic nations whose scanty population is incapable of occupying the whole'.

This principle was applied from one end of the Americas to the other, and Patagonia was no exception. To both Britain and France, Patagonia constituted a 'debatable' land, regardless of the national boundaries that supposedly incorporated it. To the Chilean and Argentinian governments, Indians were temporary occupants of the lands they had occupied for thousands of years, whose status was yet to be determined, and it was in this legal vacuum, in the last years of Rosas's rule, that the first attempts were made to establish permanent settlements in southern Patagonia and Tierra del Fuego.

* * *

In Argentina, these efforts were a direct consequence of the *Beagle*'s first two voyages and FitzRoy's and Darwin's writings. Despite the British Admiralty's lack of interest in FitzRoy's civilizing experiment in Tierra del Fuego, his attempts to establish a Christian mission at the 'end of the world' did not go unnoticed in his native land. In the early 1840s, a former British naval officer-turned-missionary named Allen Gardiner set out to reconstitute Richard Matthews' failed mission on Navarino Island. Born in Berkshire in 1794, Gardiner was a devout Anglican who first felt the missionary calling in his thirties, following the deaths of his first wife and mother. After four years in South Africa, Gardiner travelled through Bolivia and Chile as a lone missionary in 1838, where he found little enthusiasm for the Anglican faith either among the Indigenous Peoples of Araucanía or the 'Romish missionaries' who oversaw their religious instruction.

After reading FitzRoy's journal, Gardiner became convinced that Patagonia was the key to the evangelization of South America. In 1844, he founded the Patagonian Missionary Society and tried to raise money to fund a schooner that would bring the Gospel to 'the scattered tribes of Patagonia'. When these efforts fell short, Gardiner settled for two launches and dinghies. In December 1850, he and six companions were dropped off in the Beagle Channel, where they hoped to make contact with Jemmy Button. This mission rapidly unravelled when the missionaries discovered they had left their gunpowder and shot on the ship, which meant they were unable to hunt. They then lost their rations and their dinghies while trying to escape a group of Yagáns whom they assumed were hostile. When one of their launches was smashed to pieces during a storm, Gardiner and one of his companions went ashore at Spaniard Harbour, on the coast of the Isla Grande, while the other five men anchored a mile and a half away at the Cook River.

Without food, and terrified of the natives they had come to save who might have helped them, the missionaries succumbed to scurvy, starvation and exposure. Gardiner fully embraced his martyrdom. His journal is filled with prayers and expressions of gratitude for his sufferings, as well as pity for the 'poor blind heathen'. On 29 August 1851, in a final message to his wife, he celebrated his imminent death as an act of martyrdom, in which 'I am passing through the furnace, but, blessed be my heavenly Shepherd, He is with me and I shall not want ... I trust poor Fuegia and South America will not be abandoned.'[3] On 5 September, he wrote his last entry. Later that month, a rescue ship from Montevideo found the bodies of all seven missionaries, lying buried or out in the open. An 1852 editorial in the *Times* condemned 'the unutterable folly of the enterprise as it was conducted' and called for an end to future 'Patagonian missions'.[4] But Gardiner's sacrifice caught the imagination of a Victorian public enamoured with sacrifices on behalf of the faith. In 1854, the Patagonian Missionary Society launched another fundraising appeal for a floating mission ship that would be based in the Falklands/Malvinas and fill Tierra del Fuego with 'gardens, and farms and industrious villages ... [where] the church-going bell may awaken these silent forests; and round its cheerful hearth and kind teachers, the Sunday School may assemble the now joyless children of Navarin Island'.

This time, the target was reached. In 1854, the society bought the schooner *Allen Gardiner* and appointed the veteran English sailor and Arctic explorer William Parker Snow (1817–95) as its first captain. The mercurial captain quickly fell out with the mission's catechist Garland Phillips, and these disputes continued even after the mission arrived on Keppel Island in the Falklands/Malvinas in January 1855. In November that year, the *Allen Gardiner* entered the Murray Channel where Darwin and FitzRoy had said goodbye to Orundellico/Button more than two decades earlier. This time, the missionaries found Button, his two wives and their several children. In an 1861 address to the London Ethnological Society, Snow compared the 'rude, shaggy, half repulsive-looking' Button to the 'petted idol of friends here at home, [who] had been presented to royalty, and finally sent back to Fuego as a passably finished man'. Snow presented Button with FitzRoy's before-and-after sketches of the Fuegians, which 'made him laugh and look sad alternately, as the two characters he was represented in, savage and civilized, came before his eye'.[5]

In June 1858, the mission's superintendent, Reverend George Pakenham Despard, persuaded Button and his family to return to the Keppel Island settlement. For the next five months, the 'passably finished' Button and his family were given religious instruction and taught how to use soap, knives and forks and perform manual tasks such as collecting firewood, carpentry and cutting peat while the missionaries began to compile a dictionary of the Yagán language.

In November 1858, Button and his family were returned to Wulaia Cove on Navarino Island, where the missionaries built a wooden church with the reluctant assistance of Button and his fellow-Yagáns. By this time, Snow had been replaced by a new captain named Robert Fell, who transported another group of Yagáns from Wulaia to Keppel Island in January 1859. On 2 November that year, the Yagáns were returned to Tierra del Fuego accompanied by Phillips and eight other missionaries. On leaving the island, the overbearing Despard ordered a search of their bags, which uncovered a number of smuggled tools and other objects. The angry and humiliated Yagáns were taken back to Wulaia Cove, where the crew went ashore for a service in the church, leaving only the cook, Alfred Coles, on board.

Shortly afterwards, the horrified cook watched helplessly as some 300 Fuegians attacked the church and killed Phillips, the captain and six crew members with clubs, stones and spears on the beach. Coles managed to escape in a boat and was rescued by another group of Yagáns. Four months later, he was rescued by an American ship and claimed that Button had instigated the massacre. In March 1860, Button was brought to Port Stanley to answer these charges. Cross-examined by the colonial secretary, he denied any involvement in the massacre and said that it had been carried out by a group of Selk'nam from the Isla Grande. The inquest accepted this explanation, or at least could not disprove it, and Button was allowed to return to Wulaia. The *Bristol Gazette* lamented the loss of 'so many brave Englishmen' at the hands of Indians who were 'perhaps without exception, the lowest in the scale of humanity' and called on the society to abandon its 'Quixotic enterprise' in Tierra del Fuego. The British assistant undersecretary of state accused the society's 'Chief Missionary' of having 'kidnapped natives, and then kept them as forced labour', before concluding 'it is not surprising that murder should follow'.

These damning judgements failed to cool the society's ardour. In 1863, the discredited Despard returned to England with his family, leaving behind his twenty-one-year-old adopted son, Thomas Bridges (1842–98), as acting superintendent. In January 1863, the society sent another missionary team to Keppel Island under its secretary, Reverend Waite Hockin Stirling. Stirling and Bridges returned to Wulaia Cove and rebuilt the church with the help of Button and some of the Fuegians who had previously attacked it. In 1867, Stirling and Bridges established a new mission on the Isla Grande, to which they gave the name Ushuaia—the Yagán word for the area. In 1871, the renamed South American Mission Society sent a prefabricated steel and zinc house to Ushuaia as a permanent residence for Stirling and Bridges.

FitzRoy did not live to see this outcome. On 30 April 1865, he succumbed to the depression that had so often threatened to overwhelm him and cut his throat with a razor. Button also died before the Ushuaia mission was established. In 1864, Stirling returned another group of Yagáns to Wulaia Cove, where he learned that a 'malignant illness' had killed many Fuegians, including Button. According to Stirling, Button's wife came to the ship in a canoe, her face 'visibly impressed with sorrow', and 'pointing to the sky, she gave me to understand by looks, more than words, the cause of her grief, and how great it was'. Despite—or perhaps because of—his association with missionaries, Button's family did not give him a Christian burial. Instead, he was cremated and passed into history—the protagonist of an unlikely journey back and forth between different worlds that was as unlikely and astonishing in its own way as any adventure from *Gulliver's Travels*. And by that time, the Anglican mission in Tierra del Fuego had also begun to attract the attention—and the concern—of the Chilean government regarding the still-disputed sovereignty of the Strait of Magellan.

* * *

Around 2,000 miles separate Santiago from the city of Punta Arenas, on the Strait of Magellan. Even today, the plains between Puerto Natales and Punta Arenas seem endless, and you can drive for miles and miles along the edge of the sea and see nothing except a few scattered huts (*ranchos*) and sheep ranches and the occasional bare flat-topped mountain. These are the grassy, windswept flatlands that Gabriela Mistral describes in her melancholy Patagonian poem 'Desolación'—an empty land of 'immense painful sunsets', fog, sea and snow. The sparseness of the population adds to the sense of a region that is still only tenuously connected to the rest of the country. As late as the 1970s, the Pinochet dictatorship regarded the conquest of the south as a patriotic test of military mettle through the construction of the 'Carretera Austral'—the 'southern highway' that stretches 770 miles from Puerto Montt to Villa O-Higgins. But that highway has yet to overcome the natural barriers posed by lakes and forests that would connect it to the southernmost provinces of Aysén and Magallanes.

In the first half of the nineteenth century, these territories were inhabited by the Kawésqar or Alacaluf 'canoe Indians' who lived on the northern banks of the Strait of Magellan, and also by the Aónikenk or southern Tehuelche 'horse Indians', whom Darwin and FitzRoy had entertained on the *Beagle* at Bahía San Gregorio in 1834. Almost nothing was known of these peoples, and for much of the first half of the nineteenth century the Chilean government had no interest in them. Separated from the far south by the autonomous Mapuche fiefdom of Araucanía, the territories of the far Chilean south were only accessible by sea, and the government in Santiago had neither the inclination nor the resources to establish its authority any further. With the advent of the age of steam, these territories acquired a new economic and political importance. From his Peruvian exile in the 1830s, the former 'liberator-general' of independent Chile, Bernardo O'Higgins, urged his compatriots to establish a permanent presence in the Strait of Magellan. In a letter to his English friend Captain John Smith in 1836, O'Higgins proposed the establishment of a steam-powered towing service that could drag sailing ships through the straits. In another letter in 1841, O'Higgins called for ports and anchorage points to support this service, and he also stressed the 'benefits of civilization and religion' for 'the poor naked savage inhabitants of Tierra del Fuego and western Patagonia, whose miserable and wretched state is a stain on Christianity and on my country especially'. O'Higgins cited 'the interesting work by Captain FitzRoy' as 'a noble exception to the general apathy' towards the far south and its peoples.[6]

In August 1842, he urged the Chilean president Manuel Bulnes to bring 'the blessings of civilization and religion' to the 'wretched, naked and ignorant inhabitants of western Patagonia and Tierra del Fuego'. Even on his deathbed on 24 October 1842, O'Higgins exclaimed 'Magallanes, Magallanes' as a warning and an exhortation to his countrymen. The towing service never materialized, but the following year Bulnes commissioned Juan Williams Wilson (or Juan Guillermos), an English captain in the Chilean navy with Chilean nationality, to establish a permanent colony in the Strait of Magellan that would deter any potential foreign intruders.

On 22 May 1843, Williams set sail from Valparaíso in the *Ancud*, a 27-tonne schooner, accompanied by two artillery officers and five soldiers, nine sailors, a carpenter, two women and Williams' young son. The *Ancud*'s passengers also included the Prussian naturalist and geographer Bernhard Eunom Philippi (1811–52), a fervent proponent of German immigration to his adoptive country who was appointed by President Bulnes as governor of the new colony in order to encourage this process. On 21 September, the *Ancud* reached the Punta Santa Ana and raised the Chilean flag in the Strait of Magellan for the first time. Two days later, a French ship anchored in the same bay, and Williams and Philippi persuaded its captain to lower the French flag. The colonists also put up signs proclaiming 'Republica de Chile' and 'Viva Chile' to remove any remaining ambiguity over the ownership of the western straits. On 30 October, they constructed a stockade, which they called Fort Bulnes, and the following month the fort was visited by a party of Aónikenk, in the first of many trading exchanges with the settlers. A 'Treaty of Friendship and Commerce' formalized these relationships, and the Aónikenk recognized the fort and agreed to extend their protection to its members.

The establishment of the colony was regarded in Santiago as a great patriotic achievement, but few settlers were attracted to an isolated, windblown stockade with no obvious economic potential. In 1848, the colonists moved 38 miles northwards to the more sheltered bay dubbed Sandy Point by seventeenth-century English sailors, which the colonists renamed Punta Arenas. That same year, the Chilean government designated the settlement a penal colony in an attempt to boost its population. In October 1851, the first twenty convicts arrived at the colony. Within days of their arrival, Lieutenant José Miguel Cambiaso, an officer in the garrison who had been expelled from the army for bad conduct and then readmitted, led a rebellion against the new colonial governor. Cambiaso was an alcoholic and something of a psychopath. After murdering the governor and promoting himself to major general, he subjected the colony's 700-odd inhabitants to a reign of bacchanalian terror that might have been written by Joseph Conrad or filmed by Werner Herzog.

Captain Charles Brown, an American seaman taken prisoner by Cambiaso along with his crew, described his captor as a vain and capricious man with 'a beauty to his face that would have been a study for a painter' were it not for the eyes that 'revealed the evil passions hid under that fair exterior'.[7] After burning the chapel, and hoisting a red flag painted with the skull and crossbones and the motto 'conmigo no hay cuartel' (with me there is no quarter), Cambiaso's men shot, bayoneted and hanged anyone who refused to join them or obey their commander's erratic orders. According to Brown, four Indians who came to the colony to trade were 'hung by the neck to the trees, and lanced to death, their cheeks and noses being cut off' for no reason at all. In January 1852, the mutineers burned most of the colony to the ground and commandeered Brown's ship and another British ship they lured into the bay with the vague intention of escaping to Brazil. Following a request from the Chilean government, the British warship *Virago* was sent south to Punta Arenas, where it rescued forty-five passengers abandoned by Cambiaso's drunken crew in the Strait of Magellan. Cambiaso and his men were captured and taken to Valparaíso, where Cambiaso was executed and dismembered in June that year in front of 20,000 spectators.

The consequences of Cambiaso's mad rampage had not yet played out. In the same month that Cambiaso was executed, the German botanist and explorer Philippi was appointed governor of the Magallanes colony and placed in charge of the reconstruction and development of Punta Arenas. A pioneer in attracting German emigration to Chile, Philippi had originally been contracted by the Prussian government to collect natural specimens in Chile for the Natural History Museum of Berlin, and his explorations of southern Chile had induced him to settle there permanently. Philippi had joined the *Ancud* mission as a volunteer, and his knowledge of French had helped prevent a French ship from hoisting the French flag on the Strait of Magellan. In October 1852, Philippi set out on horseback with a small escort of soldiers and Indian guides with the intention of re-establishing friendly relations with the Tehuelche. On 27 October, the party was attacked by a group of Tehuelche seeking revenge for Cambiaso's outrages. Philippi and his men were killed, save for a teenage soldier who was captured and later escaped to describe what had happened. The Tehuelche had taken their revenge, and in the years that followed peaceful relations between settlers and Indians were re-established as Punta Arenas was rebuilt and continued to serve as a penal colony and a precarious outpost of Chilean sovereignty in the 'empty lands' of the far south.

9

HINTERLANDS

With the fall of Rosas, a new era began for both white Argentina and the Indigenous Peoples of the south that would ultimately bring the Argentinian army back to the territories the dictator had abandoned. In 1853, the Constitutional Convention at Santa Fe enacted a national Argentine constitution that incorporated many of the ideas of the Generation of '37. Free trade and a free press, freedom of worship, navigable rivers, state education, full political rights, unlimited European immigration—all these aspirations of Alberdi and the returned exiles found a place in the new constitution. The new dispensation also addressed Argentina's 'Indian question'. Article 67, Section 15 of the new constitution enjoined the government to 'provide for the security of the frontiers' and 'preserve the peaceful dealings with the Indians' while also calling for 'their conversion to Catholicism'.[1] National unity remained elusive on the ground, however. In September 1852, Buenos Aires seceded from the Argentine Confederation. In 1859, the army of Buenos Aires under the command of the soldier and statesman Bartolomé Mitre (1821–1906) was defeated by the confederation army at the Battle of Cepeda. Two years later, Mitre defeated confederation forces at the Battle of Pavón and became the first elected president of a national Argentine government, with Buenos Aires as its capital.

The political conflicts between the provinces also changed the balance of power between settlers and Indians in the Pampas. The day after Rosas's defeat at Caseros on 3 February 1852, 5,000 warriors from his ally Calfucurá's Salinas Grandes Confederation attacked Bahía Blanca. Mitre denounced Calfucurá's 'savage tribes'

115

and called for 'the steely argument of the sword' to be 'employed to the last degree until they are exterminated or cornered in the desert'.[2] In February 1855, Calfucurá launched another huge raid on the garrison town of Azul, in which 300 soldiers were killed and hundreds of captives and thousands of head of cattle taken. In May that year, Mitre led an expedition of 900 cavalry and infantrymen from Buenos Aires in pursuit of Calfucurá's raiders and their booty.

Mitre expected easy victories, but the expedition quickly unravelled as Mitre's army wandered the plains and sierras beyond Bahía Blanca in search of Calfucurá's raiders, while small bands of warriors carried out hit-and-run attacks on his columns and picked off stragglers. Six soldiers were beheaded and twenty knifed to death. One soldier had his throat cut and somehow survived by walking 30 miles to a nearby ranch. On 31 May, 500 warriors attacked Mitre's army in the hills known as the Sierra Chica, south of Azul, and nearly destroyed it. Forced to retreat on foot, Mitre wrote a public letter in which he blamed his defeat on the cowardice of his soldiers, 'who were so terrified of the enemy that they had fired off shots across the camp at the slightest sound, and threatened their commander with lances when he tried to contain them'.[3] Like many before him, Mitre concluded that 'the desert is unconquerable', and many of his countrymen reached similar conclusions regarding the Mapuche chief who had defeated him.

* * *

In the wake of this humiliation, the Buenos Aires press dubbed Calfucurá 'the Napoleon of the Pampas'. In June 1855, Calfucurá led another raid on Azul in which 500 soldiers and settlers were killed. In September, 2,000 Ranquels wiped out all but two members of the 150-strong garrison at San Antonio de Iraola. In October, Calfucurá lured an army under the command of General Manuel Hornos into quicksand at the Estancia San Jacinto, near the present-day city of Olavarría, killing eighteen officers and 259 soldiers, before launching a series of raids across a 341-mile radius from Junín in the north-west to Bahía Blanca in the southeast. These confrontations were on an entirely different scale to the Rosas years. For nearly two decades, Rosas's frontier commanders had distributed more than 2,000 cattle and horses to Calfucurá's confederation every month. With the breakdown of this system, the raids on white settlements increased exponentially.

At this point in history, the balance of military power in the Pampas was still relatively even. On paper, frontier armies were better armed, with firearms and occasional pieces of artillery, while their Indigenous opponents continued to use bolas, knives, machetes and the bamboo *chusa* lance—an 18-foot-long spear tipped with bayonets, scissors, swords or pieces of sharpened iron and steel. These lances were used with great dexterity by mounted horsemen, whose knowledge of the terrain gave them a significant advantage over their enemies. Operating in disciplined tactical formations under Calfucurá's skilful generalship, the Indians of the Pampas were elusive and dangerous opponents. In May 1859, Calfucurá led a 3,000-strong assault on Bahía Blanca, which was repelled by Argentine army units supported by Garibaldi's Italian Legion. Between 150 and 200 Indians were killed, whose bodies were burned in a giant pyre in the main square. Despite such losses, Calfucurá remained a constant threat to the security of Buenos Aires Province, and his military prowess was matched by his skills as an orator and diplomat that made him the dominant unquestioned *toki* (war chief) of the south. In 1856, the French adventurer Auguste Guinnard was captured by Puelche Indians and taken to Calfucurá's *tolderías*. The Frenchman briefly became Calfucurá's secretary, writing letters to the government on the cacique's behalf, and his description of 'the Grand Cacique of the Indian Federation' is one of the few first-hand accounts of the man who dominated the Pampas for more than fifty years:

> He was a man already more than a hundred years of age, though appearing to be sixty at most; his still black hair covered a vast unwrinkled forehead, which bright and scrutinizing eyes rendered highly intelligent. The entire physiognomy of this chief, though stamped with a certain dignity, nevertheless recalled the type of the Western Patagonian, to whom he owed his origin. Like them he was of high stature; he had very wide shoulders, and a protruding chest; his back was slightly bowed, his walk heavy, almost difficult, but he still enjoyed all his faculties.[4]

Despite his reputation as a war chief, Calfucurá continued to negotiate with the government in Buenos Aires. At the National History Museum in Buenos Aires, visitors can see the official seal of the Salinas Grandes Confederation with which he stamped hundreds of letters to presidents, generals, priests and frontier officials. To some extent, his raids were intended to pressure the government of Buenos Aires into honouring the agreements that Rosas had made. But in the years that followed the overthrow of the dictator, very little remained of the pro-Indian sentiments that had once led the leaders of Argentina's independence struggle to consider Indians as fellow-citizens. In 1865, Colonel José Benito Machado, a cattle rancher and commander of the southern coastal frontier district, founded a fortress bearing his name between Tandil and Bahiá Blanca on the site of present-day Tres Arroyos—encroaching still further on the boundaries of Calfucurá's confederation. In 1866, some of the largest landowners in the country founded the Sociedad Rural Argentina (Argentine Rural Society) to promote the interests of Argentina's cattle and sheep farmers in the south. Many members of 'la Rural' were bankers, politicians and public officials, and this powerful lobby group played a key role in the enactment of Law 215 in 1867 by the National Congress, which called for the government to extend the frontier to the Negro and Neuquén rivers.

This law effectively rendered Rosas's system null and void, and the response was not long in coming. In 1870, Calfucurá attacked Fort Machado on two occasions. On 23 October that year, 2,000 Mapuche warriors attacked Bahía Blanca once again. On 5 March 1872, 6,000 Mapuche and Ranquel warriors attacked settlements across Buenos Aires Province and came within 137 miles of the capital, killing 300 people and capturing 200,000 head of cattle. These raids were unlike anything taking place anywhere on the continent in this period. The Mapuche distinguished between the *tautulún*, the raid as vendetta; the *maloca*, a raid aimed at expropriating cattle and captives; and *weichán*—a defensive war aimed at halting or driving back the line of white settlement. At various times, Calfucurá's raids belonged to each of these categories and sometimes combined all of them. Yet even in the violent post-Rosas years, peace treaties were agreed in which caciques ceased raiding in exchange for rations and the acceptance of territorial boundaries.[5] These agreements

Fig. 1: Illustrations of 'Fuegians' from the voyage of the *Beagle*.

Fig. 2: *Expedition in the deserts of the South* by Calixto Tagliabúe. The painting shows Rosas on the black stallion directing operations against the 'savage Indians' in 1833.

Fig. 3: Charles Darwin and his eldest son William Erasmus Darwin, 1842.

Fig. 4: Juan Manuel de Rosas, Argentine dictator, 1829.
Portrait by Arthur Onslow.

Fig. 5: 'Waki killing a Puma', illustration from *At Home with the Patagonians* (1871) by George Chaworth Musters, showing Musters's Tehuelche hosts on a hunting expedition.

Fig. 6: *La vuelta del malón* by Ángel della Valle, 1892. Regarded at the time as Argentina's 'first genuinely national work of art', this depiction of Indian 'barbarians' returning to the Pampas from a raid was specifically painted for the 1893 World's Columbian Exposition in Chicago, to show the world what Argentina had overcome.

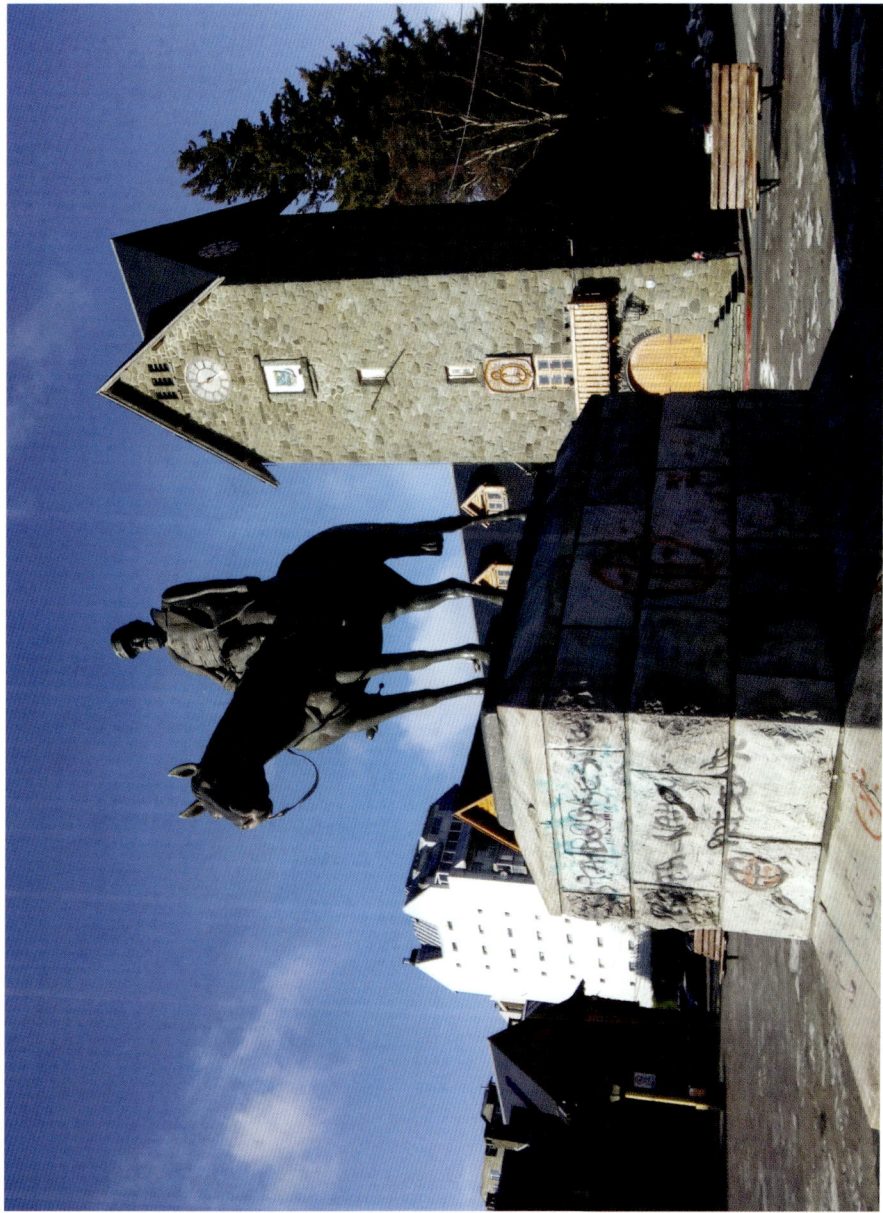

Fig. 7: Statue of General Julio Argentino Roca in Bariloche, Argentina, June 2020.

Fig. 8: *Indian Auxiliaries on Choele Choel by Antonio Pozzo, Album de Vistas Expedición al Río Negro* (1879).

were broken at various times by both sides, but they nevertheless enabled truces and periods of relative tranquillity in the frontier zone, where Indians and settlers, living in close proximity, loathed and feared each other.

* * *

The American historian Frederick Jackson Turner famously described the 'continually advancing frontier line' in the United States as 'the outer edge of the wave—the meeting point between savagery and civilization'.[6] Argentina's internal frontiers also 'advanced' during the nineteenth century. In 1784, the southern frontier line of Buenos Aires Province along the Río Salado extended only 76 miles at its southernmost point to Chascomús. By 1875, a line of forts, towns and settlements extended in a diagonal line from San Luis—492 miles west of Buenos Aires—to Carmen de Patagones on the Atlantic coast, where Darwin had begun his ride across the Pampas. The territories beyond the frontier line were named after their tribe or their individual caciques, or simply as 'various tribes of Indians', and beyond the Río Negro lay Patagonia's *campos no explorados* (unexplored territories), whose inhabitants were listed either as Tehuelches or as 'various warlike tribes'.

In theory, these forts and settlements marked the limits of state authority and the point where civilization gave way to the 'desert'. But these imagined distinctions between the civilized world of settler-colonial Argentina and its Indigenous 'barbarians' were not always as clear-cut on the ground as they seemed from a distance. The city of Azul lies some 177 miles south of Buenos Aires. At first sight, it looks like many other Pampean cities, with its rows of mostly low-rise offices and flats and single-storey buildings in the Spanish colonial style rising out of the plains. It's not the most obviously literary city, but it has fourteen bookstores in the small city centre—not bad for a population of 60,000—and a statue of Don Quixote pays tribute to Azul's reputation as the Cervantine Capital of Latin America derived from its unique collection of more than 500 editions of *Don Quixote* in a local library-cum-cultural centre.

Azul also has other less-known literary connections. The Anglo-Argentinian writer William Henry Hudson (1841–1922) and his brother did their military service here. Jorge Luis Borges's grandfather Colonel Francisco Borges Lafinur was stationed in Azul for a year and a half as a frontier commander, and his grandson made at least two visits to the city as a writer. None of this seemed likely when the city was founded by Rosas in 1833 during his march to the south. That same year, Azul's first doctor and surgeon arrived from Buenos Aires, and the following year he was sacked by Azul's commanding officer Pedro Burgos, who accused him of spending more time with his mistress than treating the sick. By 1834, no replacement had been found, and Burgos informed his superiors that he had been obliged to solicit the services of an Indian *curandero*— medicine man—to treat his soldiers.

Such were the options in this bleak outpost of civilization. 'Here, there are no pleasures, no sweetnesses, nothing to please the heart or the spirit,' lamented one member of the garrison. 'One lives far from any caress, like a parasite, with nothing to look forward to but the lance of an Indian, and nothing to look back on but continual hunger.'[7] When MacCann visited Azul in 1845, he found 'an assemblage of ranchos' with a church and a flour mill, inhabited by some 1,500 people, including British and European construction workers, who lived 'in a state of continual alarm' due to the 'Indians along the frontier.'

In 1865, the army officer, writer and politician Colonel Álvaro Gabriel Barros García (1827–92) arrived in Azul in the middle of winter to take up his new position as commander of the southern frontiers and found a garrison manned by some 400 National Guardsmen, most of whom lacked uniforms, horses, weapons and basic supplies. Azul was a frequent target of Indian raids and the point of departure—and sometimes the point of retreat—for military expeditions against the caciques of the south. The city was also a distribution point for the rations supplied by Buenos Aires to Indians who remained loyal to the government, and it acquired a permanent population of some 6,000 *indios mansos* who camped outside the town. Most of these 'tame' Indians belonged to the *tolderías* of the Catriel dynasty, founded by Rosas's ally Juan Catriel.

From the point of view of Buenos Aires, Azul was the point where civilization gave way to 'the immense regions of the unknown Pampas', as the chronicler and one-time Azul resident Antonio del Valle once put it, where only 'the echo of the savages, like the squawking of eagles, the neighing of colts, like the bellowing bull ... were felt like the echo of a hurricane storm that destroys everything in its destructive path'.[8] Violence and the threat of Indian raids were persistent features of the frontier zone. These raids generally followed a similar pattern. Small parties of raiders would ride all night towards their target, switching between horses to increase mobility so that they were in position to carry out their attacks at dawn. These assaults would usually begin by setting fire to the roof of the house in order to force its occupants into the open. The men and the elderly would then be killed with lances, knives and bolas, and the younger women and children taken as captives, along with their horses and cattle. Indigenous women sometimes accompanied the raids to guard the horses and looted domestic objects from the burning property while the men rounded up captives and animals.

Larger raids might be carried out in daylight, as hundreds or thousands of Indian warriors attacked towns and forts in full-scale assaults and drove away their supplies of cattle and horses, sometimes setting fire to the Pampas grass to cover their retreat. The authorities struggled to respond to these attacks. Without experienced *baquianos* (scouts), frontier soldiers stood little chance of catching up with their perpetrators. Even when these raiders carried off thousands of head of cattle along the wide and well-trodden *rastrilladas* that connected the Atlantic to the Andes, pursuing soldiers and militiamen could run out of food and water, lose their way or drown in quicksand. Unable to rely on the military, the wealthier landowners recruited their own private armies and turned their ranches into fortresses. But even the most well-armed rancher-militias were not always able to prevent the theft of cattle across hundreds of miles of unfenced territory. Small farmers and settlers could only defend their *ranchos* with ditches, stockades or even rows of prickly pear. Many failed and lost their lives, their property or their families to Indian raiding parties who could disappear into the emptiness of the Pampas as rapidly as they had emerged from it.

* * *

The violence and cruelty of the frontier zone was not one-sided. 'It is true that frontier settlements and ranches were ravaged by the savage,' wrote the Argentinian naturalist Francisco Moreno,

> but on the other hand how many of the latter were the old people, women and children who fell in the surprise attacks on the camps by the troops; shot and atrociously staked, victims of the soldiers who obeyed and interpreted, for good or ill, the order and gesture of a superior officer![9]

In W.H. Hudson's short story 'El Ombú', the narrator Nicandro tells the story of an 'Indian invasion on the southern frontier', in which the Azul garrison is attacked as a diversion while 'the larger number of the savages were sweeping away the cattle and horses from the country all round'. The Azul garrison commander raises an expeditionary force to pursue the Indians and eventually catches up with them. One of the expedition's participants describes how

> the savages, rushing hither and thither, trying to save themselves, were shot and speared and cut down by swords. One desire was in all our hearts, one cry on all lips—kill! kill! kill! Such a slaughter had not been known for a long time, and birds and foxes and armadillos must have grown fat on the flesh of the heathen we left for them. But we killed only the men, and few escaped; the women and children we made captive.[10]

This account was fictionalized, but the history of frontier warfare in the Pampas is filled with similar massacres and counter-massacres, which generated a legacy of hatred and suspicion on both sides. To white settlers and the Argentine government, Indians were 'foreign' invaders from Chile, savages who were incapable of civilization and bandits who lived by raiding and theft. This imagery is most famously captured in the epic historical painting *La vuelta del malón* (The return of the Indian raid, 1892), by the nineteenth-century artist Ángel Della Valle, which hangs in the National Fine Arts Museum in Buenos Aires. Originally commissioned as Argentina's contribution to the Columbian World Exhibition in Chicago in 1893, the picture depicts an Indian band returning from a night raid at dawn. The sky is grey and ominous, and the Indians are dark-skinned and mostly bare-chested; the few faces that can be seen are twisted into ferocious and barely human expressions as they brandish their lances, herd stolen horses and wave a looted crucifix and censer. But the painting's most striking feature—and the one that would have instantly attracted the attention of its nineteenth-century viewers—is the warrior with one arm round a white female captive and carrying a lance with another.

The woman is sitting sideways, and her Titian-like skin tone immediately stands out from the murky gloom. Her white dress has been pulled down around her waist, a crucifix hangs down between her bare breasts, and her head rests on her captor's shoulder. Is she a Christian martyr-maiden, resigned to her ultimate defilement at the hands of barbarians? Or has she willingly succumbed to the erotic appeal of her 'savage' captor? In 2022, the Argentinian anti-racist collective Identidad Marron filmed a clever *desmonte* (deconstruction) of Della Valle's painting inside the museum, in which a white actress and an Indigenous actor on a plastic horse reversed the central image of the civilized captive and her barbarian captor so that the blonde woman in the dress holds the lance while a smiling 'Indian' captive rests his head on her shoulder.[11]

In mocking Della Valle's racialized sexual hysteria, the video/performance also critiqued the notion of the frontier as a civilizational boundary. The taking of captives was a recurring theme in settler-colonial representations of the Indigenous Peoples of the south as inveterate raiders operating outside civilized norms.[12] There is no doubt that captive-taking was a routine practice in the nineteenth-century frontier zone. Indian raiders regarded white women and children as war booty as well as valuable commodities. Some were kept by their captors as slaves. Others were sold or returned to their communities in exchange for ransom payments, negotiated by priests or frontier commanders. The fate of these women was often tragic and horrific. Many female captives were *cuarteleras* (barracks women) who accompanied soldiers to the frontier, and the wives and daughters of settlers suffered an equally bleak fate. Some never saw their families again. Others were beaten or had the soles of their feet cut to stop them running away.[13]

Once again, such practices were not limited to the 'barbarian' inhabitants of the Argentine 'desert'. Soldiers and settlers took Indian women and children as sexual slaves and domestic servants.

Some were taken to families in Buenos Aires and other cities. Others were raped or taken by force to frontier forts and barracks to provide sexual services. There were also captives—both whites and Indians—who adopted the ways of their captors. In the 'Story of the Warrior and the Captive', Jorge Luis Borges tells the story of his English grandmother Fanny, the wife of Colonel Francisco Borges, who befriended an English captive woman from Yorkshire in 1872. According to Borges, this 'blonde Indian woman' came regularly to Junín to buy trinkets and spoke in

> a rustic English interspersed with Araucan and the pampas's dialect … A savage life could be glimpsed behind her tale: the horsehide Indian huts, the fires made of manure, the feasts of scorched meat … the naked horsemen charging the haciendas; war, polygamy, stenches, magic. An Englishwoman had lowered herself to this barbarism.

When Borges' grandmother tried to persuade her to return to civilization, the Englishwoman replied 'that she was happy, and, that night, returned to the wilderness'.[14]

* * *

Such transitions were not uncommon in a frontier population that included 'friendly' Indians, descendants of black slaves and settlers from a range of nationalities, in addition to convicts, criminals, fugitives from justice and military deserters, some of whom preferred living with Indians to the harsh discipline of garrison duty. Spanish-speaking captives who could read and write were sometimes employed as interpreters and translators by their captors. Some of them became the secretaries of Pampean caciques and frontier-diplomats. Santiago Avendaño was captured by Ranquels at the age of eight. During his seven-year captivity, Avendaño learned the Ranquel language, and following his return to 'civilization' he went on to become an intermediary between the Argentine government and the Pampean caciques.

Although Indians and settlers were competitors for land and resources, they were also economically interdependent. Darwin's own descriptions of the *indios mansos* at Carmen de Patagones and

Bahía Blanca are a testament to trading relationships that benefitted whites and Indigenous People. Such was the economic importance of Patagones as a trading entrepôt that the Tehuelche cacique Casimiro Biguá was given the rank and pay of a lieutenant-colonel in the Argentine army in return for an agreement to protect the settlement in the event of an invasion by Calfucurá. The Tehuelche often complained they were cheated by white traders at Patagones, but Tehuelche bands continued to travel there every year to trade from the Aónikenk or southern Tehuelche territories adjoining the Strait of Magellan.

Trading often took place at the *pulpería* (bar-cum-grocery store)—the central social institution of the Argentine frontier, where Indians came to sell or exchange hides, ostrich feathers and other hand-made products for coveted 'white' goods such as tobacco, sugar, yerba maté and alcohol. In 1900, Conrad's friend, the writer and seasoned Argentina traveller Robert Cunninghame Graham described the typical *pulpería*, where customers could find 'ponchos from Leeds, ready-made calzoncillos, alpargatas, figs, sardines, raisins, bread ... saddle-cloths', as well as an array of drinks that included 'vermuth, absinthe, square-faced gin, Carlon, and Vino Seco'.[15] In times of war, the *pulpería* could become a mini-fortress where settlers defended themselves against Indian attacks. But such was the importance of this institution to Indigenous society that such defences were not always required. The journalist Estanislao Zeballos (1854–1923) described a *pulpería* near the Río Colorado whose owner lived beyond the frontier line in 'such good harmony' with the local Indians for nearly thirty years that his cattle and house were left untouched even in times of war.

However fragile and impermanent, 'good harmony' was also part of the frontier. Records from the Azul historical archive in 1857 show that the local council staged a special Mass and days of celebration to mark the signing of a peace treaty with the Pampas Indians. If there was war in the frontier zone, peace and coexistence were also possible, even in the 'gateway to the desert', and beyond these outposts of civilization, visitors continued to travel further south into territories that many of them had first discovered through Darwin's writings.

10

TRAVELLERS AND SETTLERS

Between 1869 and 1870, the former British naval commander George Chaworth Musters (1841–79) rode 1,400 miles from the Magellan Strait to Carmen de Patagones on the Río Negro as a guest of the Tehuelche. A veteran of the Crimean War, Musters was an eccentric and unusually intrepid traveller. He arrived at Punta Arenas wearing a guanaco cloak and equipped with a lasso and bolas with the intention of visiting a territory he had discovered through Darwin and FitzRoy. In *At Home with the Patagonians: A Year's Wanderings over Untrodden Ground from the Straits of Magellan to the Rio Negro* (1871), Musters paid tribute to 'Mr. Darwin's work on South America, as well as FitzRoy's admirable Narrative of the Voyage of the Beagle' as an inspiration for his travels in the company of a Tehuelche band from the Santa Cruz River who agreed to take him northwards to Carmen de Patagones. Musters was one of a number of travellers who wrote about Patagonia in the second half of the nineteenth century. Guillermo Cox's *Viaje en las rejiones septenrionales de la Patagonia 1862–1863* (Journey in the southern regions of Patagonia, 1863); Guinnard's captive narrative *Trois ans d'esclavage chez les Patagons* (Three years of slavery among the Patagonians, 1861); and Julius Beerbohm's *Wanderings in Patagonia* (1879) all continued the Patagonian adventure narrative that Darwin and FitzRoy had effectively begun.

Some came as tourists, others as naturalists, adventurers, explorers and settlers. Their books were not only read for entertainment. In 1862, the Chilean adventurer Cox travelled across the Andes on behalf of the Chilean government to investigate the possibility of a sailing route from northern Patagonia to the

Atlantic. Equipped with chronometers, barometers and a guitar, flute and flageolet, Cox's team attempted to navigate the Limay River and had to be rescued by Indians when their raft capsized. A translation of Cox's paper on his attempts to 'open a new route across the continent' was presented to the Royal Geographical Society by the former British consul in Buenos Aires, Sir Woodbine Parish, in the presence of the then-Admiral FitzRoy, who offered his meteorological observations on its contents.[1]

In 1863, the Welsh Emigration Society published a Welsh-language *Handbook on the Welsh Colony*, which recommended Patagonia as the ideal location for the establishment of a Welsh colony, drawing on FitzRoy's and Darwin's writings. Like Darwin and FitzRoy before them, these Patagonian travel books contained ethnological observations, maps and descriptions of peoples and landscapes that were studied by politicians and geographical and anthropological institutes in Chile, Argentina and Europe. On the one hand, their journeys were a testament to Patagonia's accessibility and proximity to the wider world. At the same time, their writings were instrumental in bringing Patagonia and its peoples to the world's attention, in ways that both confirmed and sometimes contradicted the assumptions of those who had been there before them.

* * *

If Musters was inspired by FitzRoy and Darwin to visit Patagonia, his descriptions of its Indigenous Peoples were more sympathetic than his predecessors. Musters lived with his Tehuelche hosts for the best part of a year. He slept in their tents and participated in their feasts, hunts, card and ball games and drunken parties. Musters was at pains to give his readers a different idea of the Tehuelches 'than that which ... has usually been assigned to them' and insisted that his hosts 'do not deserve the epithets of ferocious savages, brigands of the desert, &c. They are kindly, good tempered, impulsive children of nature, taking great likes or dislikes, becoming firm friends of equally confirmed enemies.'[2] He rejected the stereotypes of female slavery, domestic violence and the absence of familial attachments that Darwin and so many others had viewed as a hallmark of savage societies and described the freedom of Tehuelche women to contract marriages, flirt and leave their husbands. 'The finest trait, perhaps, in their character,' he wrote, 'is their love for their wives and children; matrimonial disputes are rare, and wife-beating unknown; and the intense grief with which the loss of a wife is mourned is certainly not "civilised", for the widower will destroy all his stock and burn all his possessions.'

Musters was so enraptured by his journey across Patagonia that he continued to sleep under a blanket in his garden until his marriage in 1873. He and his wife continued to travel until his death in 1879, on the eve of a voyage to Mozambique. His maps and descriptions became an important resource for Argentinian politicians and soldiers and attracted other English travellers to Patagonia, not all of whom saw its peoples in the same positive terms. In 1877, the engineer Julius Beerbohm, the elder half-brother of the famous caricaturist Max Beerbohm, accompanied a group of engineers on a survey expedition of the territory between Port Desire and Santa Cruz. Beerbohm estimated that 3,000 Tehuelche lived in Patagonia, and he believed that these numbers were declining with such rapidity that 'it is to be feared that in a comparatively short period they will have disappeared altogether from the face of the earth, and survive only in memory as a sad illustration of the remorseless law of the non-survival of the unfitted'.[3]

Other European travellers reached similar conclusions. In 1876–7, the travel writer Baroness Annie Brassey sailed around the world on a ship called the *Sunbeam* carrying forty-three family members, friends and crew, two dogs, three birds and a Persian kitten. In *A Voyage in the 'Sunbeam': Our Home on the Ocean for Eleven Months* (1878), Brassey describes her attempts to see authentic 'Patagonians' during a visit to Punta Arenas. Unable to find any Tehuelche, she and her party were taken by the British vice-consul to see 'three Fuegian females' who had been rescued in a canoe after seeking refuge 'from some kind of cruelty or oppression' and were recovering from their injuries at the house of the colony's chief medical officer. Although the women appeared 'cheerful and happy', Brassey's party were told 'they are not likely to live long. After the free life and the exposure to which they have been accustomed, civilization—in the shape of clothing and hot houses—almost always killed them.'

These women were 'certainly not handsome', but Brassey was nevertheless surprised to find 'they are by no means so repulsive as I had expected from the descriptions of Cook, Dampier, Darwin, and more recent travellers'.[4]

Other visitors found what they had expected to see. In 1879, Beerbohm returned to Punta Arenas with a party that included the poet and travel writer Lady Florence Dixie. In *Across Patagonia* (1881), Dixie described the horrified responses of her friends and acquaintances to her intended journey to a land where she was likely to be 'eaten up by cannibals', and her own attraction to Patagonia '[p]recisely because it was an outlandish place and so far away … Palled for the moment with civilisation and its surroundings, I wanted to escape somewhere, where I might be as far removed from them as possible.'[5] Dixie was one of the few female writers in a Patagonian travel genre dominated by men, but her observations of Patagonia's Indigenous Peoples were trite, superficial and contemptuous. At Bahía San Gregorio, she described her first 'real Patagonian Indian' as

> a singularly unprepossessing object, and, for the sake of his race, we hoped an unfavourable specimen of it. His dirty brown face, of which the principal feature was a pair of sharp black eyes, was half-hidden by tangled masses of unkempt hair, held together by a handkerchief tied over his forehead, and his burly body was enveloped in a greasy guanaco-capa, considerably the worse for wear.

Subsequent encounters did nothing to dispel her disdain for the Tehuelche men she regarded as lazy cannibals and domestic tyrants. Like Beerbohm, Dixie saw the Tehuelche as 'a race that is fast approaching extinction' and predicted that their over-consumption of alcohol would turn them into 'a pack of impoverished, dirty, thieving ragamuffins'. It was a measure of Darwin's status as the doyen of Patagonian travel writers that Lady Florence sent him a copy of her own book—a gift that was graciously accepted.

One of the more eccentric visitors to Patagonia was Orélie-Antoine de Tounens (1825–78), the French lawyer and self-styled 'King of Patagonia and Araucanía'. Tounens was inspired to become an explorer by his readings of La Pérouse, Cook and other explorers, and the epic sixteenth-century Spanish poem *La Araucana* instilled in him a desire to visit the region. In 1858, he sailed to Chile and crossed the Araucanian frontier with the assistance of a well-known Mapuche chieftain. For reasons that have never been made clear, Orélie persuaded the Mapuche to declare independence and crown him king. In 1860, Tounens granted himself the right to create 'nobles' across 'one single nation under a Government of Constitutional Monarchy' that extended from the Strait of Magellan to Argentinian Patagonia. Soon afterwards, he was arrested by the Chilean army, imprisoned as a madman and deported to France. In 1869, he returned to Patagonia and tried in vain to reconstruct his 'kingdom' in Argentinian Patagonia.

Once again, these efforts failed, and Tounens ended up in Buenos Aires with an intestinal blockage and finally returned to Europe. The 'Kingdom of Araucanía and Patagonia' still exists, at least in theory. Tounens had no descendants, but some of his countrymen have laid claim to his legacy. The late Jean Raspail, the author of the racist novel *The Camp of the Saints* (1973), also wrote a novel about Tounens. On the basis of his research, and his own visits to Patagonia and Tierra del Fuego, Raspail transformed his own home into a 'consulate-general of Patagonia' as an expression of the 'tenderness, irony, pride, and melancholy' that he considered to be essential hallmarks of Patagonian identity.[6]

* * *

Like so many nineteenth-century Europeans who came to Patagonia, Tournens was attracted, at least initially, by what he read about the region, and these travellers' tales also played an indirect role in the establishment of one of the more improbable European settlements in the Argentine south. Today, visitors to the Andean town of Trevelin, near Esquel, in the Patagonian province of Chubut, will immediately notice the large sculpture of a winged dragon perched on the roof of the wooden town hall, the trilingual signs in Welsh, Spanish and Mapudungun, the plethora of names like Evans, Roberts, Morgan and Thomas, and the cafés with their equally incongruous offer of *té gales*—Welsh tea. Founded by Welsh colonists in 1891, Trevelin marks the westernmost limits of the chain of Welsh settlements established in Patagonia in the second half of the nineteenth century.

The decision to establish a Welsh colony in Patagonia followed many years of discussion among devout, patriotic Welshmen about a viable location for a colony where the Welsh language and religious faith could be protected. In 1861, the Liverpool Emigration Society was founded to bring this project to fruition. Two years later, the society published a Welsh-language handbook, which quoted selectively from FitzRoy and Darwin, leaving out the more negative pronouncements of both writers, such as Darwin's description of the Patagonian plains as 'useless to mankind ... the curse of sterility is on the land'.[7] The committee chose the mouth of the Chubut River as the ideal location for the colony after reading 'the report of Admiral FitzRoy (at the time when Darwin was with him)'. The society sought and received the cooperation and assistance of the Argentinian government, which saw a Welsh settlement in Patagonia as a means of neutralizing Chilean territorial claims east of the Andes. On 28 May 1865, a 447-ton tea clipper, the *Mimosa*, left Liverpool with 153 Welsh settlers and arrived on 28 July at New Bay (Puerto Madryn), where they were greeted by the society's two delegates Lewis Jones and Edwyn Cynrig Roberts.

The colonists were shocked to find that the fertile paradise they had been led to expect was, as one settler described it to his relatives, 'an almost infinite plain, without trees or rocks, only different types of grasses and bushes, some reaching above head-height, so it is very difficult to move around and only the luckiest comes out of it with his skin intact'. The settlers promptly moved to a new location that they named Rawson—a tribute to Guillermo Rawson, the Argentinian minister of the interior. Most of these settlers were Welsh mining families, few of whom had any knowledge of farming or agriculture. Stranded in a remote territory, some 154 miles south of the nearest settlement at Carmen de Patagones, they struggled to overcome the poor soil, floods and a succession of bad harvests.

Their survival was partly due to the supplies they received from the Argentinian government and the assistance of the local Tehuelche. The colonists were initially so fearful of Indian attacks that they trained an armed militia to protect themselves, but their amicable interactions with the Tehuelche were among the more agreeable surprises that Patagonia had to offer. The Tehuelche taught the Welsh how to hunt rhea and guanaco and earn money by exporting feathers and skins to Buenos Aires. Such assistance was not entirely motivated by altruism, as local Tehuelche chiefs were paid 300,000 pesos by the Argentinian government in exchange for protecting the settlement. In December 1865, the Tehuelche cacique Antonio sent a letter to the settlement leader Lewis Jones, which assured him:

> Our plains have plenty of guanacos and plenty of ostriches. We are never in want of food ... I and my people are contented to see you colonize on the Chupat, for we shall have a nearer place to go in order to trade, without the necessity of going to Patagones, where they steal our horses and where pulperos [storekeepers] rob and cheat us. If you treat us well ... and if your traders do not cheat us, we shall always negotiate with you.[8]

The Welsh accepted this offer and forged lasting ties of friendship with the Tehuelche. Some of them married Indian women and persuaded some of the Tehuelche to refrain from hunting on Sundays out of respect for the Welsh Sabbath. Such was the respect in which the Welsh were held, according to historian Glyn Williams, that the Tehuelche distinguished their Welsh *hermanos* (brothers) from the hated settlers they called *cristianos* (Christians)—a distinction that ignored the fact that the Welsh were also Christians. As they prospered, new settlements were built westwards along the length of the Chubut River and into the Andean foothills. By the late 1870s, the colonists were producing wheat, butter, flour, alfalfa and wool for export, building irrigation systems along the Chubut River and importing flour mills and agricultural machinery. And throughout the westward expansion of Y Wladfa (The Colony), their relationships with the Indigenous Peoples of Patagonia were conducted on an entirely different basis from the brutal confrontation that was unfolding further north.

11

OF APES AND MEN

The establishment of the Welsh colony also coincided with a period in which Darwin had begun to look back on his encounters with Patagonian 'savages' as he turned his attention to the ongoing debates over human origins and the 'varieties of man'. In January 1862, Reverend Charles Kingsley wrote to Darwin on the subject of the 'great gulf between the quadrumana & man; & the absence of any record of species intermediate between man & the ape'. The author of *The Water Babies* (1863) had once described the Irish poor as 'white chimpanzees' in a letter to his wife. Now, he speculated that the abundance of half-animal 'mythological personages' found in many different cultures might constitute the ancestral memory of some vanished 'semi-human race' or 'creatures intermediate between man & the ape' in the distant past, and he suggested to Darwin that their disappearance might have been the result of 'simple natural selection, before the superior white race'.[1] In reply, Darwin described the 'genealogy of man' as 'a grand & almost awful question', which

> is not so awful & difficult to me, as it seems to be most, partly from familiarity & partly, I think, from having seen a good many Barbarians. I declare the thought, when I first saw in T. del Fuego a naked painted, shivering hideous savage, that my ancestors must have been somewhat similar beings, was at that time as revolting to me, nay more revolting than my present belief that an incomparably more remote ancestor was a hairy beast.[2]

Darwin did not pronounce judgement on Kingsley's mythological speculations, but he agreed with 'what you say about the higher races of men, when high enough, replacing & clearing off the lower races. In 500 years how the Anglo-Saxon race will have spread & exterminated whole nations; & in consequence how much the Human race, viewed as a unit, will have risen in rank.'

This exchange took place less than three years after the publication of *On the Origin of Species by Means of Natural Selection: Or the Preservation of Favoured Races in the Struggle for Life* (1859), in which Darwin finally revealed his theory of evolution to the world. This book transformed Darwin into the most influential and controversial scientist of the Victorian era, the progenitor of a new vision of life and nature that was simultaneously world-shattering, heretical and liberating. Human beings were largely absent from its pages, but Darwin still found space for the 'barbarians of Tierra del Fuego' in a discussion of domestic animals, in which he repeated William Low's claims that the Fuegian 'barbarians' were so attached to their dogs that they resorted to 'killing and devouring their old women' rather than their animals 'during famines and other accidents, to which savages are so liable'.[3] Few of Darwin's readers would have disputed such claims, at a time when the 'genealogy of man' had become one of the burning scientific mysteries of the day.

* * *

The publication of *Origin* coincided with a new phase in the history of scientific racism in which discussions about racial difference and the 'genealogy of man' increasingly focused on biological, pseudo-biological and physiological factors rather than soil, diet, climate or environment. Robert Knox's *The Races of Men* (1850); Arthur de Gobineau's *Essai sur l'inégalité des races humaines* (Essay on the inequality of the human races, 1853–5); Josiah Nott and George Gliddon's gargantuan 800-page tome *Types of Mankind* (1854); George Fitzhugh's *Sociology for the South* (1854)—all these books reflected an epistemological shift in the scientific understanding of race, in which human development was understood as the outcome of a life-and-death struggle between the 'lower' and 'higher' races. To some extent, these writers applied a form of evolutionary theory *avant la lettre* to the subject of race in their assumption that white Europeans or Caucasians had been victorious in this struggle, and the certainty of this victory was often accompanied by the belief that the world's 'lower races' were condemned to disappearance through their defeat.

To the anatomist and former army surgeon Knox, history was 'the old tragedy' of 'the fair races of men against the dark races; the strong against the feeble; the united against those who knew not how to place even a sentinel; the progressists against those who stood still—who could not or would not progress'.[4] Writing five years before *Origin*, the pro-slavery sociologist Fitzhugh saw the 'vegetable, animal and human kingdoms' locked in 'a constant conflict, war, or race of competition, the result of which is, that the weaker or less healthy genera, species and individuals are continually displaced and exterminated by the stronger and more hardy. It is a means by which some contend Nature is perfecting her own work.'[5] For Nott and Gliddon, '[n]o two distinctly marked races can dwell together on equal terms. Some races, moreover, appear destined to live and prosper for a time, until the destroying race comes, which is to exterminate and supplant them'—a process Fitzhugh saw in the spectacle of 'the aborigines of America ... fading away before the exotic races of Europe'.[6]

Such predictions often returned to the 'savages' of the New World. According to Nott and Gliddon, American Indians were so 'incapable of development' that they were condemned to the 'earthly destiny' of extinction, in which, 'after a few generations, the red men will be gathered to the tombs of their forefathers'. The entomologist Augustus Grote described the generic Indian as a child 'who has not passed the mental state of our own children'.[7] In *Ancient Society* (1877), the American anthropologist and ethnologist Lewis Henry Morgan placed Indians at the 'zero of human society' with 'no help of elevation'.[8] Long before the BAAS recognized anthropology as a distinct branch of science in 1884, the 'science of man' had become an essential prism through which the world's Indigenous Peoples were studied, analysed and categorized. Edward Tylor, John Lubbock, Lewis Henry Morgan, Walter Bagehot, John McLennan, Augustus Pitt-Rivers—all these writers made significant contributions to the anthropological debates of the nineteenth century, and many of them were concerned with the same questions Darwin and Kingsley discussed.[9] What were the stages that marked humanity's evolutionary ascent? Were the world's 'savages' mere 'survivals' from a phase of evolution that the civilized races of the world had superseded? Or did they represent a form of degeneration or atavism? In *The Origins of Civilisation and the Primitive Condition of Man* (1870), Darwin's friend and neighbour, the banker, politician and polymath Sir John Lubbock (1834–1913) argued that the 'knowledge of modern savages and their modes of life enables us more accurately to picture, and more vividly to conceive the manners and customs of our ancestors in bygone ages'.[10]

Such knowledge, Lubbock argued, was of 'a peculiar importance to an empire such as ours, comprising races in every stage of civilisation yet attained by man'. Lubbock was a strong critic of the 'degeneration' school represented by writers such as Richard Whately, the archbishop of Dublin. For Lubbock, the 'savage traces' visible in his own times were the residual legacy of an 'ancient barbarism' that civilization had long since superseded. The 'true savage', Lubbock argued, was a tragic anachronism, a 'slave to his own wants, his own passions', constantly driven by hunger, 'always suspicious, always in danger, always on the watch. He can depend on no one, and no one can depend upon him.' Civilization, by contrast, was a process of continual improvement and perfection that heralded 'even greater human happiness in the future'.[11]

* * *

The logical corollary of civilization's march towards perfection was the disappearance of the races and peoples that were not capable of improvement. In a review of Richard Lee's *The Extinction of Races* in 1864, James Hunt (1833–69), the racist co-founder of the London Anthropological Society, described 'the rapid disappearance of aboriginal tribes before the advance of civilisation' as 'one of the many remarkable incidents of the present age'. Between 'the white and the coloured populations there are not even degrees of civilisation', Hunt proclaimed: 'The man who now wanders free through the unknown wilds of Australia represents nothing. Not only has he not advanced in moral development since the first formation of his species, but he has actually retrograded.' Faced with this 'illustration of humanity in its crudest form, shrinking and passing away before a phase of humanity enlightened with intelligence, and endowed with vast intellectual superiority', it was only a short time before 'the surface of the earth will be utterly altered; whole races, which now rules supreme over immense tracks, will have passed away for ever'.[12]

In *Scenes and Studies from Savage Life* (1868), the Vancouver magistrate and author Gilbert Malcolm Sproat attributed the 'decline through colonial contact' among the First Peoples of Canada to a kind of civilizational shock, in which 'the savage' was 'confused and his faculties surprised and stunned by the presence of machinery, and the active labour of civilized men', to the point that he 'distrusted himself, his old habits and traditions, and shrank away despondent and discouraged' into a state of stunned paralysis.[13] Sproat exhorted his compatriots to avoid 'harsh measures' against these doomed Indians and hoped that '[a] clear view of the impending extinction of the inferior people would probably rather stimulate English settlers to acts of justice and humanity towards them'. If such peoples were condemned to disappear by natural laws, then their disappearance could be regarded with equanimity. For the Scottish botanist and explorer Robert Brown, '[t]he disappearance of wild races before the civilised is, for the greater part, as explicable as the destruction of wild animals before civilised sportsmen'.[14]

With the publication of *On the Origin of Species*, these 'doomed race' narratives increasingly acquired a biological tinge. Lubbock hailed his neighbour's 'great principle of natural selection' as an essential instrument of civilizational progress.[15] In *The Struggle for Existence* (1888), Darwin's intellectual paladin Thomas Henry Huxley attributed the disappearance of prehistoric societies to a biological struggle in which 'the weakest and stupidest went to the wall, while the toughest and shrewdest, those who were best fitted to cope with their circumstances, but not the best in any other sense, survived'.[16] Even the socialist Alfred Russel Wallace, Darwin's co-discoverer of natural selection, invoked the 'great law of the "survival of the preservation of favoured races in the struggle for life"' in an 1864 lecture to the London Anthropological Society as an explanation for 'the inevitable extinction of all those low and mentally undeveloped populations with which Europeans come in contact'.[17]

In *The Malay Archipelago* (1869), Wallace compared the eclipse of 'savage man' in the 'struggle for existence' to 'the weeds of Europe [that] overrun North America and Australia, extinguishing native populations by the inherent vigor of their organization, and by their greater capacity for existence and multiplication'. In *The Synthetic Philosophy* (1896), Herbert Spencer argued that 'the savage's slow ascent in the scale of unilinear evolution made him entirely insignificant in the race struggle with the nations of large-brained peoples'.[18] Commenting on the 'vital competition of Mr Darwin' in *Anthropology* (1878), the French physical anthropologist Paul Topinard described 'the fact of races inferior in the struggle becoming extinct' as an 'altogether natural' process that 'exhibit[s] to us series of generations, strata of more and more perfect races succeeding and replacing each other'.[19]

These grand narratives of the rise and fall of races coincided with the highwater mark of European colonial and imperial conquest in the second half of the nineteenth century. The British invasion of Zululand and occupation of Burma, Egypt and Sudan; the Indian Mutiny and the imposition of British direct rule; the 'Scramble for Africa'; the final defeat of the American Plains Indians; French military expeditions in southern Vietnam and the Mekong Delta—all these episodes were part of what seemed to be a seemingly irresistible process of European expansionism that was easily interpreted as a confirmation of racial superiority. And it was against this background that Darwin prepared to make his own contributions to the ongoing debates he had only referred to obliquely in *On the Origin of Species*.

* * *

Darwin prepared to enter these debates with his characteristic thoroughness. Within months of the publication of *On the Origin of Species*, he sent a letter to Bridges at the Patagonian Missionary Society mission on Keppel Island, which included the following questions about the Fuegians at the mission:

Do they blush? And at what sort of things? Is it chiefly or most commonly in relation to personal appearance, or in relation to women?
When out of spirits or dejected do they turn down the corners of the mouth?
Do they ever shrug their shoulders to show that they are incapable of doing or understanding anything?[20]

In his letter, Darwin explained that '[a]ny information on the manner of expression of [sic] countenance of any emotion in savages would be curious, and I believe is a subject, which has been wholly overlooked'. Bridges replied in detail to Darwin's questions. Yes, the Fuegians blushed, 'chiefly in regard to women'. They also frowned, shrugged their shoulders, cried and hurt themselves when angry. Over the next decade, Darwin sent his 'Queries about Expression' to colonial officials, settlers, travellers and missionaries across the world, and these correspondents sometimes offered unsolicited views of their own. Darwin cannot be blamed for this: few other sources were available for the kind of information he required. But these correspondents were not necessarily the most dispassionate and objective observers. In July 1867, he received a response to his questionnaire from a Cape Colony farmer and amateur naturalist named James Philip Mansel Weale, who also enclosed another response from a missionary-educated Xhosa interpreter and constable named Christian Gaika. Darwin wrote back, describing the reply from 'the brother of a Kaffir [Xhosa] chief' as 'a truly wonderful fact in the progress of civilization', to which Weale sourly described Gaika's ability to write as 'an error into which most people in England fall', since 'the longer a Kafir has been on a Mission Station the worse servant he is'.[21]

In 1869, the chemist and mineralogist David Forbes answered Darwin's questions about the Aymara Indians of Bolivia and Peru in a letter that claimed that Spanish conquistadores had larger penises than Indians and attributed this superior 'furniture' to their military success. One of Darwin's correspondents may have been Juan Manuel de Rosas. On 27 June 1863, Darwin's son William Erasmus Darwin wrote to his father from Southampton, where he had taken up a position with the Southampton and Hampshire Bank. 'I thought the simplest plan was to go and speak to Old Rosas himself, so I caught him yesterday,' he wrote. 'I tried to explain that you had met him last year and spoke to him, and I think he understood. At all events he took your paper and read it carefully through making me read each word, and then he said he would write the answer and give it to his servant for me.'[22]

Rosas had only recently moved permanently to a 300- to 400-acre farm in the village of Swaythling, just outside Southampton, because he was no longer able to afford the rent on his townhouse. The former *estanciero* who had once owned hundreds of thousands of head of cattle struggled to manage a farm with a few dozen sheep and dairy cows, and the entrepreneurial skills that had once made him one of Argentina's richest men had deserted him in rural Hampshire. Though he was able to pay the rent, his farm never became a profitable enterprise, even though the *Hampshire Advertiser* later remembered him

> constantly riding about the grounds, and his greatest happiness seemed to be to sit on his horse and give orders to those employed. His love of command was so great that no one was allowed to speak a word except by way of acknowledgment of orders given or in answer to questions.

As Rosas sank deeper into genteel rural poverty, he brooded over his downfall, wrote letters to his supporters in Argentina and compiled a Ranquel–Spanish glossary.

In 1864, the Argentinian diplomat and newspaper editor Nicolás Antonio Calvo found the former dictator 'strong and vigorous. He says that he sleeps beneath a corridor that he showed to us; that he is poor; that he has saved many papers but no money.' These papers covered 'diverse branches of human knowledge; on natural law; medical science and other subjects' that Rosas wanted published after his death.[23] There is no record of the paper Darwin sent to the general, nor is it known whether Rosas ever wrote back to him. But the most likely explanation for his son's visit is that Darwin wanted to consult Rosas in connection with the research that formed the basis for *The Descent of Man, and Selection in Relation to Sex* (1871)—a sprawling combination of natural history, anthropology, ethnology, political science and moral philosophy, in which Darwin finally asked in public the questions he had pored over for so many years.

* * *

The 'savages' of Patagonia and Tierra del Fuego appear repeatedly in Darwin's discussion of 'whether man, like every other species, is descended from some preexisting form'. Once again, the Fuegians were presented to his readers as the embodiment of the world's 'lowest savages'.[24] Where he had once pondered on the gulf between the Fuegians and Sir Isaac Newton, Darwin now compared 'the man described by the old navigator Byron, who dashed his child on the rocks for dropping a basket of sea-urchins,' to the penal reformer John Howard and the anti-slavery campaigner Thomas Clarkson in order to trace 'the steps by which some semi-human creature has been gradually raised to the rank of man in his most perfect state'. Darwin came down firmly on the side of Lubbock in his conclusion that 'all civilized nations were once barbarous'. He also rejected the 'polygenists' who regarded races as distinct species. 'I was constantly struck,' he wrote, 'whilst living with the Fuegians on board the "Beagle," with the many little traits of character, showing how similar their minds were to ours.' Darwin cited the ability of FitzRoy's Fuegian captives to 'talk a little English' as evidence that they 'resembled us in disposition'.

In a discussion of the difference between instinct and moral choice, Darwin recalled the story of the three Indians captured and tortured during Rosas's Desert Campaign, who 'preferred being shot, one after the other, to betraying the plans of their companions in war' as one of the 'many instances [that] have been recorded of barbarians, destitute of any feeling of general benevolence towards mankind, and not guided by any religious motive, who have deliberately as prisoners sacrificed their lives, rather than betray their comrades; and surely their conduct ought to be considered as moral'. If savages were 'moral', they were also subject to the same Malthusian checks that nature imposed on flora and fauna. In a discussion of 'the Extinction of the Races of Man', he argued that 'extinction follows chiefly from the competition of tribe with tribe, and race with race' in a contest that 'is soon settled by war, slaughter, cannibalism, slavery, and absorption'. In these prehistorical struggles between tribes and races, Darwin hypothesized, the 'weaker tribe' had either been 'swept away' or experienced a gradual decline until such tribes became 'extinct'.

Like many of his peers, Darwin observed the same dynamic unfolding in his own era.[25] 'When civilized nations come into contact with barbarians, the struggle is short,' he wrote, 'except where a deadly climate gives its aid to the native race.' Darwin cited the decline of the Ottoman Empire as an example of this process and reminded his readers how 'a few centuries ago Europe feared the inroads of Eastern barbarians, now, any such fear would be ridiculous'. Elsewhere, in a discussion of the evolutionary 'breaks' or gaps between different versions of the same species, he predicted that '[a]t some future period, not very distant as measured by centuries, the civilised races of man will almost certainly exterminate and replace throughout the world the savage races' in much the same way as earlier iterations of elephants and apes.

Once again, Darwin evoked the concept of 'extermination' that he had previously applied to plants and megafauna to describe the disappearance of the world's 'savage races'—a process that he suggested was a consequence of natural selection. These emphatic assertions raised more questions than they appeared to answer. What were the evolutionary factors that had enabled the 'civilized races of man' to triumph over the 'Eastern barbarians'? How had 'civilized man' transcended his lowly ancestry? Darwin's answers to these questions were often convoluted and contradictory. On the one hand, he argued that '[w]ith savages, the weak in body or mind are soon eliminated; and those that survive commonly exhibit a vigorous state of health'. At the same time, he also observed that '[w]e civilised men ... do our utmost to check the process of elimination' by building hospitals, asylums and carrying out vaccination programmes, all of which ensured that 'the weak members of civilised societies propagate their kind'.

If such practices were weakening 'the race of man' through their altruism and humanitarianism, then why were the 'civilized races' so successful? Borrowing from his cousin, the eugenicist Francis Galton, Darwin noted the 'downward tendency' in his own country by which 'the reckless, degraded, and often vicious members of society, tend to increase at a quicker rate than the provident and generally virtuous members'. He noted with approval that 'the intemperate suffer from a high rate of mortality, and the extremely profligate leave few offspring'. If this winnowing process did not operate quickly enough to 'prevent the reckless, the vicious, and otherwise inferior members of society from increasing at a quicker rate than the better class of men', then it was possible that 'the nation will retrograde, as has occurred too often in the history of the world'.

Darwin's meaning is often occluded by sweeping generalizations and a persistent slippage between nebulous and undefined terms such as savages, races and nations. In his closing arguments, the 'savages' of Tierra del Fuego appear in a striking passage in which he once again recalls 'the astonishment which I felt on first seeing a party of Fuegians on a wild and broken shore':

> These men were absolutely naked and bedaubed with paint, their long hair was tangled, their mouths frothed with excitement, and their expression was wild, startled, and distrustful. They possessed hardly any arts, and like wild animals lived on what they could catch; they had no government, and were merciless to everyone not of their own small tribe. He who has seen a savage in his native land will not feel much shame, if forced to acknowledge that the blood of some more humble creature flows in his veins. For my own part I would as soon be descended from that heroic little monkey, who braved his dreaded enemy in order to save the life of his keeper; or from that old baboon, who, descending from the mountains, carried away in triumph his young comrade from a crowd of astonished dogs—as from a savage who delights to torture his enemies, offers up bloody sacrifices, practises infanticide without remorse, treats his wives like slaves, knows no decency, and is haunted by the grossest superstitions.[26]

At that time, neither Darwin nor anyone else knew whether the peoples of Tierra del Fuego were 'merciless' towards each other, or whether they tortured their enemies, practised infanticide or were 'haunted' by 'superstitions'. The 'zookeeper's tale' referred to a story Darwin had been told at the London Zoological Gardens in which a monkey that a zookeeper kept as a pet attacked a baboon that had bitten its owner. The 'old baboon' came from the German ornithologist and hunter Alfred Brehm's *Brehms Thirlaben* (Brehm's life of animals, 1896),[27] which described how a Hamadryas baboon had rescued a younger baboon from a pack of hunting dogs.

On the one hand, Darwin suggested Fuegian 'savages' had less-developed moral instincts than apes and baboons. At the same time, the evocation of these savages and their 'grossest superstitions' had a different rhetorical function. If Darwin had been able to overcome his own shame and disgust at the prospect that he was descended from the 'frothing' sub-human Fuegians, then civilized Victorian man, 'with all his noble qualities' and 'god-like intellect', need not feel any shame at 'the indelible stamp of his lowly origin [from apes]'. Having ascended to 'the summit of the organic scale', Darwin suggested, his readers could look forward to 'a still higher destiny in the distant future'. With these rousing affirmations, Darwin reaffirmed the civilizational 'march of improvement' he had celebrated at the end of his *Beagle* journal in order to make evolution palatable to a Victorian readership that still believed humanity was a divine creation. And by this time, his readers could also be found in the countries where he had first observed the differences between civilization and savagery.

12

THE BONE COLLECTORS

The first full-length Spanish translation of *The Descent of Man* was not published until 1876, followed by *On the Origin of Species* the following year, while the *Journal of Researches* did not appear in Spanish until 1921. But long before these books were translated, Argentinian and Chilean scientists, politicians and intellectuals had become familiar with their contents through foreign editions, partial translations and discussions in journals and newspapers. In 1839, the Chilean polymath, philosopher and founder of the University of Santiago, Andrés Bello, translated extracts from Darwin's and FitzRoy's journals from the *Edinburgh Review* and the *Journal of the Royal Geographical Society* in the journal *El Araucano*, which provided the Chilean government and reading public with some of their first glimpses of the far south and the peoples who lived there. These readings found a receptive audience. As early as 1830, the Chilean government recruited the French botanist Claude Gay to undertake scientific expeditions across the country and establish a natural history collection in Santiago. In the years that followed, European savants such as the Polish geologist Ignacy Domeyko and the Philippi brothers Bernhard and Rudolf were recruited for similar purposes. Long before the publication of *Origin*, the Chilean elite saw science as the hallmark of a modern civilized state and an essential instrument for the exploration and exploitation of the country's natural resources.

Argentina's scientific institutions had developed more slowly. Despite the assistance he had given to Darwin in 1833, and the surveyors and astronomers who had accompanied his Desert Campaign, Rosas had little interest in science beyond its practical utility. When Darwin's contact in Luján, the army physician and naturalist Francisco Javier Muñiz, donated his extensive collection of paleontological specimens to Rosas's government with a view to establishing a museum of natural history, Rosas gave the collection to the French admiral Jean Henri Joseph Dupotet, who sent them to Paris. Rosas's opponents had a very different conception of science. In a pamphlet published in Montevideo in 1844, the exiled Argentinian writer and journalist Florencio Varela included the 'annihilation of every germ of morality, civilization and intellectual advancement' among Rosas's crimes and denounced the 'modern Nero' for his lack of support for Argentina's 'literary, scientific and humane institutions'.[1]

Varela was assassinated in 1848, probably on Rosas's orders, and it was not until the overthrow of the dictator that his successors began to emulate Chile's example. In 1862, the Prussian naturalist Dr Hermann Burmeister (1807–92) was appointed director of the new Museo Público (Public Museum) of Buenos Aires. This appointment was followed by a plethora of new institutions: the Sociedad Científica Argentina (National Scientific Institute) in 1872; the Sociedad Zoológica Argentina (Argentinian Zoological Society) in Córdoba in 1874; the Museo Antropológico y Arqueológico (Anthropological and Archaeological Museum) in 1877; and the Oficina Topográfico Militar (Military Topographic Office) in 1879, which became the Instituto Nacional Geográfica (National Geographic Institute) in 1901.

In both Chile and Argentina, Darwinism provoked the same fierce scientific, intellectual and theological debates as it did in other parts of the world.[2] At the same time, Darwin's associations with Patagonia and Tierra del Fuego were a source of patriotic pride in both countries. And in bringing these territories to the attention of the international scientific community, Darwin's early writings and his subsequent fame encouraged a generation of Argentinian naturalists to follow in his footsteps to the south.

* * *

One of the first Argentinians—perhaps the first—to read *On the Origin of Species* was the Anglo-Argentine writer and naturalist

William Henry Hudson. The son of American parents with English or Irish heritage, Hudson grew up on a farm in Quilmes, on the outskirts of Buenos Aires, in the Rosas years. Hudson's father was a staunch *rosista*, who kept portraits of the dictator and his wife in his house, and his son was constantly aware of the dictator's looming shadow. During a boyhood visit to Buenos Aires, he once saw Rosas's 'court jester' Don Eusebio marching past in a general's uniform with an escort of twelve sword-carrying soldiers. These jesters were intended for Rosas's amusement only, and Don Eusebio's procession was watched by a crowd of unsmiling spectators who knew that any laughter or mockery on their part was punishable by arrest or execution. Hudson later recalled the 'dull roar from distant big guns' during the Battle of Monte Caseros in 1852 and the sight of a bloody patch on the grass outside his farm where Rosas's retreating soldiers had cut the throat of one of their own officers.[3]

Despite this background of violence, Hudson looked back fondly on the childhood consolations of living close to 'a great wilderness, waterless and overgrown with thorns', where he would fall asleep to the sound of snakes hissing under the floorboards beneath his bed. In 1859, his older brother Edwin returned from a visit to England with various books, including *On the Origin of Species*. Hudson was initially unmoved. Following a second reading, however, he 'insensibly and inevitably' became an evolutionist, albeit 'never wholly satisfied with natural selection as the only and sufficient explanation of the change in the forms of life'. On completing his military service in 1865, Hudson approached Burmeister at the Public Museum in Buenos Aires to offer his services as a naturalist. With Burmeister's support, Hudson began sending bird specimens to the Smithsonian Institute and the Zoological Society of London. He also began to write essays, field reports and letters, some of which were published in the Zoological Society's *Proceedings*. In one of these reports, Hudson accused Darwin of having made a 'careless' mistake in describing the *carpintero* woodpecker as a treeless bird during what he dismissively called Darwin's 'rapid ride across the Pampas'.

Darwin replied courteously, but the insolent criticism from this unschooled maverick clearly irritated him, to the point that he damningly referred to Hudson in the sixth edition of *Origin* in 1872 as a 'strong disbeliever in evolution'. In 1874, Hudson emigrated to England, inspired by the writings of the eighteenth-century parson-naturalist Gilbert White. In a letter home, he described his first sight of Southampton, with its 'wide clean macadam streets, grand old elm and horse-chestnut trees—parks covered with velvety turf' and 'the pretty orchard and green fields of the modest thatched cottage of Rosas' that he passed in Swaythling.[4] Thus began a difficult journey from poverty and obscurity to popular recognition as a writer and co-founder of the Royal Society for the Protection of Birds. In his worst moments, alone and almost penniless in London, Hudson often pined for the wide-open spaces of his Argentine homeland.

In *Idle Days in Patagonia* (1893), he referred to his 'intense longing to visit this solitary wilderness, resting far off in its primitive and desolate place, untouched by man, remote from civilization' that inspired him to spend a year in and around Carmen de Patagones in 1870–1. This journey was partly inspired by Darwin's writings, and Hudson's meditations on Patagonia's Indigenous Peoples contained a notably Darwinian tint. The typical Indian, he wrote, lived in a 'state of intense watchfulness, or alertness rather, with suspension of the higher intellectual faculties, [which] represented the mental state of the pure savage. He thinks little, reasons little, having a surer guide in his instinct.' Like Darwin before him, Hudson asked how 'men living in a state of nature' had existed 'for thousands of years in a state of pure barbarism, living from hand to mouth' when 'a little foresight ... would be sufficient to make their condition immeasurably better'. The answer, he concluded, lay in the featureless Pampean plains, which kept their inhabitants trapped in an 'instinctive state of the human mind, when the higher faculties appear to be non-existent'.[5] Like so many of his peers, Hudson believed that the evolutionary stasis of America's 'savages' condemned them to destruction:

> Here, as in North America, contact with a superior race has debased them and ensured their destruction. Some of their wild blood will continue to flow in the veins of those who have taken their place; but as a race they will be blotted out from earth,

as utterly extinct in a few decades as the mound-makers of the Mississippi valley, and the races that built the forest-grown cities of Yucatan and Central America.[6]

In addition to bird skins, Hudson also searched Indigenous graveyards in search of arrowheads and Indigenous skulls and bone fragments, which he brought back with him to Buenos Aires. Hudson cited these 'weapons and fragments' as evidence that 'the mind was not wholly dormant' even among lost Patagonian races who were 'slowly progressing to a higher condition'. He was not the first, or the last, European to bring back human remains from Patagonia, to cater for what had already become an international market in Indigenous body parts.

* * *

Nineteenth-century racial science covered a wide range of disciplines, from anatomy, phrenology, zoology and medicine to anthropology and physical anthropology, all of which shared a common interest in the world's 'savage' peoples—an interest fuelled by the belief that such peoples were disappearing. In a 1908 lecture at the University of Liverpool, the Scottish folklorist and social anthropologist James George Frazer described 'the savage' as 'a human document, a record of man's efforts to raise himself above the level of the beast', and predicted that 'in another quarter of a century probably there will be little or nothing of the old savage life to record. The savage, such as we may still see him, will then be as extinct as the dodo.' In these circumstances, Frazer urged British universities and the government to 'secure without delay full and scientific reports of these perishing or changing peoples, to take permanent copies, so to say, of these precious monuments before they are destroyed'.[7]

Such 'documents' might consist of 'objects of natural history, such as casts of living humans or photographs', which the French Imperial Museum of Natural History included in an 1860 instruction handbook for travellers and colonial officials. But the scope of nineteenth-century racial science was not limited to living 'savages'. In an era in which 'race' was to some extent a hypothesis or set of hypotheses seeking confirmation, scientists also sought to acquire physical evidence of racial differences and human genealogy. At times, such evidence was close to hand. In 1814, a Khoekhoe woman named Sarah Baartman was brought to Europe as a slave from the Eastern Cape by an English doctor, where she was exhibited in London and Paris by an animal trainer as a 'phenomenon of nature' called the 'Hottentot Venus'. Baartman was considered such a rarity that following her death in 1815, her body was dissected by Cuvier, who examined her skull and eye sockets, the shape of her nasal cavities and her labia and buttocks. In a detailed autopsy report to the French Academy of Medicine, Cuvier compared Baartman's 'inferior head' and buttocks to an ape as possible evidence that 'Negroes and Hottentots' were subject to 'this cruel law that seems to have condemned to eternal inferiority those races with depressed and compressed skulls'.[8]

Baartman's skeleton, genitalia and brain were exhibited in the Musée de l'Homme in Paris, where her body cast and skeleton were exhibited until 1974. But scientific institutions and museums were also obliged to look further afield for osteological remains and other physical specimens. When Darwin studied medicine at Edinburgh University between 1825 and 1827, his anatomy teacher John Barclay's dissecting room contained rows of skulls of Aborigines, Maoris, Caribs and other examples of 'barbarous and animal man' ascending to an Anglo-Saxon skull that the phrenologist George Combe deemed to represent mankind in his 'highest elevation ... as a philosopher and an enlightened moralist'.[9] Such specimens were not easily acquired. Cuvier wrote a manual for European travellers who were likely to witness or take part in battles with savages, which instructed such travellers on how to obtain and transport their body parts. These collectors were advised to 'carefully note all that relates to the individual from whom the cadaver came' before boiling these cadavers in a solution of soda or caustic potash to remove their flesh, after which they could be bagged and labelled. Wherever possible, Cuvier advised, these bodies should be brought back to Europe with the flesh still intact so that they could then be soaked and dried in order to preserve their facial forms without attracting insects.

In European and American laboratories, physical anthropologists, phrenologists and craniologists pored over the differences between longer 'dolichocephalic' and shorter 'brachycephalic' skulls in an attempt to determine which race they belonged to or where they might have come from. Anatomists dissected African and Native American bodies in order to measure the weight of their brains, the shape of their skulls or the dimensions of the skull cavity. Criminologists such as Cesare Lombroso collected Indigenous skulls in order to trace the connections between savages and animals. By the second half of the nineteenth century, the museums, laboratories and 'bone rooms' in Europe and North America had generated a global market for Indigenous body parts. The 'father of physical anthropology', Pierre Paul Broca (1824–80), measured 180,000 skulls in the course of his career, while the Hunterian Museum of the Royal College of Surgeons accumulated some 3,000 human skulls from different parts of the British Empire. In 1864, the American Military Museum transferred 3,761 mostly Indigenous remains to the Smithsonian Institution's Division of Physical Anthropology.

These objects were acquired on battlegrounds, the sites of massacres, in Native American graveyards and in previously 'undiscovered' territories, such as Patagonia. In his diary, Darwin identified Patagonia as an evolutionary necropolis that contained the fossils of extinct 'monsters', and the vast territories of the far south also provided a potentially rich field of investigation for scientists and collectors concerned with the 'varieties of man'. During the first *Beagle* voyage, FitzRoy brought the pickled body of a dead Fuegian in a barrel back to England. In July 1834, Darwin discovered an Indigenous cemetery at Puerto San Julián and recruited 'a party of officers ... to ransack the Indian grave in hopes of finding some antiquarian remains'. In the second half of the nineteenth century, European collectors converged on Patagonia and the Pampas in search of the skulls and skeletal remains of the recently dead or more ancient remains. Others were obtained through direct solicitations to Argentinian scientists and military commanders. Colonel Francisco Borges donated three Indian skulls to the English doctor Joseph Barnard Davis, who included them in his *Thesaurus Craniorum: Catalogue of the Skulls of the Various Races of Men* (1867). Borges also ordered an army surgeon to collect Indigenous skulls from battlefields, some of which were donated to an Italian doctor in Buenos Aires and ultimately found their way to the Italian National Museum of Anthropology and Ethnology in Florence. In 1896, the French author and aviator Count Henri de La Vaulx spent fourteen months living with Tehuelche tribes in Chubut Province in Patagonia, where he acquired ninety-six human skulls and ten skeletons.

These excavations took little account of Indigenous sensibilities and customs regarding the treatment of the dead. On one occasion, La Vaulx discovered that two Tehuelche Indians had died in a nearby encampment two months before his arrival. La Vaulx located and excavated the graves and spent two days and nights dissecting the bodies and removing their flesh. 'This useful gift, but little appreciated by the Indians, obliged me to break relations with them,' he wrote, 'and only through presents was I able to pacify them.' La Vaulx asked his readers: 'In the end what does it matter if a Tehuelche sleeps in a hole in Patagonia or showcase in a museum?'[10]

Many Europeans regarded dead 'savages' in much the same terms. Bridges gave skeletons of his deceased Yagán parishioners to a French scientific expedition that passed through Tierra del Fuego in 1883. When two Yagáns drowned after drinking with the crew of the ship *La Romanche* during the same expedition, their bodies were preserved in formaldehyde and taken back to Paris, where they were exhibited in the Musée d'Ethnographie du Trocadéro. The zoologist, palaeontologist and fervent Darwinist Florentino Ameghino (1853–1911) supplied both animal fossils and human remains to European collectors, including a complete skeleton, which he sold to a former Italian taxidermist from the Museum of Milan in 1872.

Argentinian naturalists also travelled to Patagonia in search of Indigenous remains for their own collections and institutions. The most famous of these collectors was the explorer, geographer, educationalist, museum director and self-taught anthropologist Francisco Pascasio Moreno (1852–1919). Moreno was the only scientist of his generation to achieve the status of a national hero and the honorific title of 'perito' (specialist or expert), in recognition of his role in the 1902 Patagonian boundary negotiations with Chile.

At the museum in Bariloche that bears his name, his watch, pistol, glasses and other possessions are exhibited like relics of a secular saint, along with maps of the territories he acquired for Argentina. Born into a wealthy English-Argentinian family from Buenos Aires, Moreno later recalled his childhood readings of Marco Polo, David Livingstone and the Franklin expedition, which instilled in him a 'profound admiration for the martyrs of science, and a lively desire to follow, in a more modest sphere, of such daring endeavours'.[11]

Moreno's father was crucial to these aspirations. As the director of the Buenos Aires Province Bank, secretary of the Buenos Aires Commerce Stock Market and a future parliamentarian, his friends included Sarmiento and Burmeister, both of whom took a personal interest in the budding naturalist's career. By the age of fifteen, Moreno and his brother had accumulated an impressive collection of fossils, Indigenous bones, skulls and artifacts, which they kept in the Moreno family home in Buenos Aires. In 1872, Moreno followed Hudson's footsteps to the Río Negro, where he excavated Indigenous graveyards at the British ranches near Carmen de Patagones that Hudson had recommended to him. These excavations were intended to uncover the 'logic of physical and moral evolution' that connected stone age man to his own times, and Moreno believed that it was possible to trace this 'march to perfection' both in the 'extinguished generations that time had buried in the Patagonian maritime littoral' and in 'the tribes that succeeded them in the possession of the land'.[12]

Moreno was also keenly aware of the interest of European scientists in acquiring such specimens. On returning from his expedition, he sent a Tehuelche skull to the Paris School of Anthropology, where Topinard used it to demonstrate the difference between 'the modern Tehuelche' and a more ancient type of 'autochthonous American'. Broca described Moreno as 'a man brimming with such youth and ardor' that his work might become 'as valuable to the study of the races of South America as [Samuel] Morton's has been'.[13] With Broca's support, Moreno published an account of his expedition in the prestigious *Revue d'Anthropologie*. This was no mean feat for an untrained amateur naturalist in his early twenties without a university degree, and these successes encouraged Moreno to undertake further expeditions to Patagonia in search of the 'general burial place of all the American races during their forced migrations to the extreme south of the American continent'.[14]

Moreno's family connections helped him to secure his government's backing for these explorations. In 1874, he took part in an official expedition to the Santa Cruz River, accompanied by the Latvian/Baltic German naturalist Karl Berg, during which he obtained what he called a 'harvest' of eighty ancient Indigenous skulls. In 1879, he led an Exploratory Commission of the Southern Territories, during which he was taken prisoner by Valentín Sayhueque, the chief of the Tehuelche/Mapcuche confederation that occupied the 'Land of the Wild Apples' between the Neuquén and Limay rivers and the Andean foothills. Sentenced to death for espionage, Moreno managed to escape the night before his execution, and this adventure added to his reputation as the intrepid scientist-adventurer. Moreno met many of the caciques of the Pampas and northern Patagonia, some of whom acted as his guides and hosts. In *Viaje á la Patagonia austral* (Journey to southern Patagonia, 1879), he praised the 'generous instincts' of his Tehuelche hosts and insisted that 'the pure Indian is not the evil one who sacks the frontiers, often encouraged by third parties who call themselves Christians'.

Moreno described himself as 'the last traveller' to have observed the 'independent Indian and lord of the Pampas and summits, with no more laws than those imposed by his limited necessities'. This romanticism was accompanied by a relentless and obsessive search for Indigenous remains, which recognized no ethical or moral boundaries. In a letter to his father in April 1875, he described his attempts to obtain the skulls and remains of deceased members of the Catriel dynasty in Azul:

> Although I doubt I can obtain the number of skulls I had wished for, I am certain that by tomorrow I will have 7 ... I expect it will not be too long before I acquire the remains of the entire Catriel family ... I already have the skull of the illustrious Cipriano, and the whole skeleton of his wife Margarita. Now it appears that his younger brother Marcelino, the leader of the present uprising, hasn't long to live, either.[15]

The skull of the 'illustrious Cipriano'—the principal cacique of the Catriel dynasty at the time—was a particularly prized addition to his collection. 'I examined it a while ago, but even after a bit of cleaning, it continues to stink,' he told his father. 'It will travel with me to the Tandil, as I am unwilling to separate myself from this jewel, for which I am greatly envied.' Not surprisingly, Moreno acquired a morbid reputation among Patagonia's Indigenous Peoples, as a warlock or a *gualicho* (evil spirit). In 1875, he attempted to persuade Sam Slick, the amiable son of the Tehuelche chief Casimiro, who had acted as Moreno's guide, to pose for photographs. Moreno also tried to measure Slick's skull, but he refused on the grounds that Moreno 'wanted his head'. The young Tehuelche's fears were not without foundation. On learning that Slick had been killed in a drunken brawl in 1876, Moreno immediately set out to find his corpse. As he put it: 'I enquired into the location of his grave, and, on a moonlit night, exhumed his cadaver, whose skeleton is now preserved in the Buenos Aires Anthropological Museum. This sacrilege was committed in the service of the osteological study of the Tehuelche.'[16]

* * *

Such acquisitions, Moreno believed, would serve the interests of science by shedding light on human evolution, and he was also aware that they raised Argentina's scientific profile—and his own. By 1877, he had acquired '400 Indigenous skulls of ancient races' from the 'ancient Patagonian necropolis' and other regions of the republic, which he donated to the Anthropological and Archaeological Museum at the Teatro Colón, in Buenos Aires. Photographs from Moreno's skull collection were exhibited at the Argentinian pavilion at the 1878 World Exhibition in Paris, inscribed with the place where they were found or identified by their ethnic groups, and Moreno subsequently donated an album of some fifty cranial photographs to the Societé d'Anthropologie de Paris.

Moreno's closest competitor as an Indigenous skull collector was the lawyer, ethnographer, politician and founder of the Argentinian Geographical Institute Estanislao Zeballos. The author of various books on Patagonia and its Indigenous Peoples, Zeballos is the embodiment of the cultural historian Jens Andermann's depiction of nineteenth-century Argentinian anthropology as a 'pillaging science'.[17] As a journalist, he frequently accompanied military expeditions to the south, which he used as an opportunity to add to his collections. A photograph shows Zeballos wearing a poncho and seated outside a *rancho*, flanked by soldiers, with three skulls arranged on a shelf. When an army officer expressed disapproval at the skulls Zeballos carried in his travel bag, he replied:

> My dear lieutenant ... if Civilization has urged you to twist and persecute their race and conquer their lands, so Science urges that I serve her by bringing the skulls of Indians to the museums and laboratories. Barbarism is cursed, and not even the remains of its dead will remain in the desert.[18]

Few of Zeballos's contemporaries expressed the symbiosis between racial science and military conquest with such cynical relish. And at a time when the struggle with the Indigenous Peoples of the Pampas had turned decisively in favour of the state, the 'desert' provided Moreno, Zeballos and their fellow-collectors with an abundant harvest of skulls.

13

THE GREAT WALL OF ARGENTINA

In March 1870, President Sarmiento sent an army officer named Colonel Lucio Mansilla on a diplomatic expedition to the Ranquel fiefdom of Leubucó to conclude a peace treaty with the Ranquel chief Mariano Rosas. This expedition took place at the end of Argentina's victory as a member of the Argentina–Brazil–Uruguay Triple Alliance in the 1865–70 war against Paraguay—the bloodiest war in South American history—in which Mansilla had served with distinction before his appointment as commander of the southern frontier forces in Córdoba Province. There was an unlikely family connection between the frontier emissary and the Ranquel cacique. Mansilla was Rosas's nephew, and Mariano Rosas was Panguitruz Guor, 'Hunter of Lions'—the young Ranquel captive whom Rosas had brought up as his godson before the boy escaped and returned to his tribe. Accompanied by two Franciscan friars in a demonstration of his peaceful intentions, Mansilla set out on an eighteen-day journey through *tierra adentro*, which he chronicled in a series of articles for the Buenos Aires newspaper *La Tribuna*.

These dispatches were later published as *A Visit to the Ranquel Indians* (1870), one of the key texts of nineteenth-century Argentinian frontier literature. Mansilla was a curious, patronizing but not unsympathetic observer of Indigenous Argentina. Though he had no reservations about calling the Ranquels 'barbarians' even in their presence, he told his readers that 'the barbarians are in no way inferior to the best of republics'. He opposed what he called 'this constant cry for the extermination of the barbarians' on the very un-nineteenth-century grounds that 'the facts that have been observed concerning the physical constitution of races are too sparse to enable us to draw general consequences from them when it is a matter of condemning entire populations to death or to barbarism'. His ethnographic observations are interspersed with philosophical reflections, vignettes and descriptions of some of the prominent Indian caciques of the period, from Vicente 'Catrunau' ('Hunter of Jaguars') Pincén, Mariano Rosas's brother Epumer and Manuel Baigorria to the forty-five-year-old Mariano Rosas himself. Mansilla found the object of his expedition dressed 'handsomely, but without luxury. He received me in a Crimean shirt, golden brown trimmed with black braid, a silk kerchief at his throat, *chiripá* of English poncho cloth, fringed pantaloons, calf-skin boots, broad belt with four silver buttons, and a hat of fine beaver with a broad red ribbon.'

The two men had various discussions, both in public and in private. In one of these public debates, Rosas asked Mansilla in front of his fellow-caciques: 'How can the land not belong to us, if we were born in it?' Mansilla explained that government forces had occupied lands near the Río Quinto 'for the greater security of the frontier'. Though 'these lands did not yet belong to the Christians', he explained, they would eventually 'belong to one man, or two, or more, when the government sells them for stockbreeding, for sowing wheat and maize'. Mansilla described the exchange that followed:

> 'You ask me, by what right do we collect land?'
> 'I ask all of you—by what right do you invade us to collect stock?'
> 'It's not the same thing!' several broke in. 'We don't know how to work; nobody has taught us how, as the Christians were taught. We are poor; we have to go on raiding to live.'
> 'But you steal what is not yours,' I said to them, 'for the cows, the horses, the mares, the sheep you bring away are not yours.'
> 'And you Christians,' they answered me, 'take away our land.'[1]

This discussion touches on the very different conceptions of land ownership at the heart of the confrontation between the Argentinian government and its Indigenous Peoples, and it also captures the suspicion with which each side often viewed the other. Though Mansilla persuaded Rosas to accept Sarmiento's new territorial boundaries, the treaty was not approved by the Argentinian Congress, which was increasingly unwilling to engage in *negocio pacífico* with the Pampean caciques.

* * *

This intransigence was the result of various factors. At the beginning of the nineteenth century, the population of the Argentinian territories was estimated at just over 600,000. In 1869, the first national census counted 1,887,490 members of the Argentine Republic. Between 1868 and 1874, 3,000 miles of telegraph were laid in Argentina, and in 1870 the first transoceanic telegraph cable connected Argentina to Europe. In 1876, the first *frigorífico* (refrigerated ship) completed a return journey from France to Buenos Aires, and the following year the French refrigerated ship *Le Frigorifique* left Buenos Aires for France with a cargo of frozen meat.

These demographic and technological developments transformed Argentina's southern 'deserts' into increasingly desirable acquisitions. In the aftermath of Paraguay's defeat, a triumphant government began to bring its full resources to bear on its 'internal' frontiers. 'For the Indians of America I feel an invincible repugnance with no possible remedy' wrote Sarmiento in the newspaper *El Nacional* in 1876, only six years after sending Mansilla to negotiate with the Ranquels. Sarmiento railed against the 'lousy Indians whom I would order to be strung up whenever they reappear ... Incapable of progress. Their extermination is providential and useful, sublime and great. They should be exterminated without even sparing the little one, who already feels the instinctive hatred towards civilized man.' In 1878, the Buenos Aires newspaper *La Prensa* described Argentina's conflict with its Indigenous Peoples as

> a contest of races in which the Indigenous life carries on itself the tremendous anathema of its disappearance, written in the name of civilization. Let us morally destroy that race, let us annihilate their economy and their political organization, let us eliminate their tribes and if necessary divide the families.[2]

In both Chile and Argentina, the 'Indian question' was increasingly described in quasi-scientific language as a racialized war against

161

the autonomous Mapuche or 'Araucanian' territories south of the Río Biobío. In 1855, *El Correo del Sur* condemned the 'shameful' situation in which a 'civilized nation such as Chile' was obliged to respect treaties with the Mapuche, whom it described as a 'weak people, barbarous and without money'. In an 1859 editorial, the Valparaíso newspaper *El Mercurio* called on the Chilean army to occupy Araucanía 'in the name of civilization' and described 'the Indian' as 'entirely uncivilizable; nature has wasted everything on his body, while his intelligence has remained at the same level as wild animals, whose qualities he possesses to a high degree, never having had a moral emotion'.[3] In a debate in the Chilean Chamber of Deputies in August 1860, Senator Antonio Concha echoed these calls for the military conquest of Araucanía on the grounds that 'our flag will never flutter in Indian territory if we continue to pretend that we can subjugate and civilize those barbarians by means of colonies, commerce, industry and missions'.[4] In 1862, the Chilean army crossed the Biobío River and began the two-decade-long military campaigns known to official Chilean history as the 'Pacification of Araucanía', which the Mapuche remember as the *wingka malon* (white raid). And in Argentina, the calls for the subjugation and elimination of Sarmiento's 'lice-ridden Indians' of the south were conditional on the defeat of the most powerful Mapuche cacique of the Pampas.

* * *

Even before the end of the Paraguayan War, settlers and soldiers had already begun to bear down on the strategic triangle of Salinas Grandes–Carhué–Choele Choel that formed the heartland of Calfucurá's Salinas Grandes Confederation. In September 1868, Calfucurá wrote to Colonel Álvaro Barros, the commander of the southern frontiers, to inform him that two Argentinian colonels had been reconnoitring Choele Choel. 'They tell me that forces have already arrived at Choele Choel and that they are coming to make war on me,' Calfucurá warned. 'I have also sent my commission to my brother Reuquecurá asking for people and forces; but if they withdraw from Choele Choel, nothing will happen and we will be fine.'[5]

On this occasion, the warning was heeded, and the reconnaissance mission withdrawn, but tensions and skirmishes continued. In 1869, Calfucurá's son, Bernardo Namuncurá, warned Barros that his brother had arrived at Choele Choel with 3,500 'lances' and urged him to keep his father informed of the government's intentions so that these forces would not be required. Barros was not unsympathetic to such requests. Like Mansilla, he was critical of the 'intelligent and enlightened men' who called for the 'extermination of the barbarians' and rejected the depictions of the Pampean Indians as inveterate raiders.[6] Barros frequently denounced the corrupt mismanagement of the government's rations system, which he argued was a major contributing factor to Indian raids on white settlements, but these views were increasingly at odds with those of his government. In 1870, the Sociedad Rural called for a 'radical change' of policy in response to the 'continual invasions and depredations that the wild Indians make against our frontiers'. The following year, the society presented a memorandum to the government, which claimed that 'more than 5,000 Chilean Indians' had entered the south and called for the 'expulsion of the barbarians to the south of the Río Negro'. The minister of war replied that such a campaign was already in preparation and that it had to be planned with care to avoid the risk that 'the arms of civilization retreat as on other occasions from the lance of barbarism'.

Calfucurá was not unaware of these developments. Like Mariano Rosas, he kept abreast of political developments in the capital through his secretaries. The Napoleon of the Pampas often expressed his desire for peace in the many letters he dictated to his secretaries, but he was always ready to resort to war when his demands were not met. In 1872, 3,500 Ranquel and Mapuche warriors from his confederation attacked settlements across the south of Buenos Aires Province in response to the arrest of three of Calfucurá's Indian allies by the Azul garrison commander, Colonel Ignacio Rivas. On 6 March, Rivas left Azul with a punitive expedition of 600 soldiers, national guards and 1,000 Indians under the command of the loyalist cacique Cipriano Catriel. With a house and bank account in Azul, and his own personal carriage, the corpulent Cipriano was not the most formidable ally, but he played a crucial role in the events that followed.

Rivas originally intended to cut off Calfucurá's line of retreat southwards, but when the recently raised fort at San Carlos, north-west of Azul, sent out a call for assistance, his men joined the garrison's defenders instead. On 8 March, Calfucurá's forces attacked the fort in a crescent formation. Calfucurá also took the unusual decision to order hundreds of his men to dismount in a demonstration that Indians could fight on foot as well as Christians. The fort's defenders included battle-hardened soldiers from the Paraguayan War, some of whom were armed with the new Remington breech-loading repeater rifles. According to the official version of the battle, Cipriano posted sharpshooters behind his own lines with orders to shoot any of his men who refused to fight against fellow-Indians. Artillery and rifle fire tore gaps in the Mapuche–Ranquel formations before Rivas led a charge directly through the centre of Calfucurá's advancing columns. Soldiers and Indians stabbed, cut and battered each other with sabres, knives and bolas before Calfucurá ordered his warriors to retreat, leaving 300 of his men dead on the battlefield.

The Battle of San Carlos was the most serious setback Calfucurá had experienced in nearly half a century, and it proved to be the final battle of his long career. On 4 June 1873, he died surrounded by his sons, wives and closest lieutenants after warning his heirs on his deathbed: 'Don't surrender Carhué to the white man.' In November that same year, Cipriano allied himself with his former enemy Bartolomé Mitre, who refused to accept the electoral victory of his rival Nicolás Avellaneda in the presidential elections. Cipriano was captured before he was able to join Mitre's forces and handed back to his own tribe for trial and punishment. Sentenced to death for treason by his brothers Juan and Marcelino and the other leaders of his tribe, he and his translator, the captive-turned-frontier diplomat Santiago Avendaño, were repeatedly struck with lances before being decapitated and thrown into a ditch. As a result of this tragedy, Cipriano's head became Moreno's prized possession, presumably through a private transaction with Cipriano's brother.

* * *

Calfucurá's death was a turning point in the struggle between Argentina and the Pampean caciques. At the time of his death, his combined forces may have amounted to 20,000 *indios de pelea* (fighting Indians), but their leaders lacked the military and diplomatic skills that had enabled Calfucurá to maintain his inter-tribal alliances for nearly half a century. On 12 October 1874, the recently invested president Avellaneda appointed the lawyer and politician Adolfo Alsina as minister of war and the navy, and the following year Alsina announced a new 'plan for the south', which entailed the extension of the frontier line 45 miles southwards, with a chain of forts, watchtowers and settlements linked, for the first time, by telegraph to the Ministry of War in Buenos Aires. Alsina also proposed another strategic innovation: a *zanja* or ditch, 2.6 metres wide and 1.75 metres deep, fortified with watchtowers and barbed wire, that would eventually reach across the entire length of the new frontier from Bahía Blanca in the south to San Rafael in the present-day province of Mendoza.

Alsina's 'Great Wall of China' in the Pampas was widely mocked. Some of his critics described it as a form of defeatism. Others criticized its cost. One of the most prominent opponents of Alsina's 'wall' was the overall commander of Argentina's frontier forces, General Julio Argentino Roca (1843–1914). In December 1875, Alsina asked Roca for his 'frank opinion' on his proposals. In a forthright reply, Roca declared his opposition to 'defensive war' as a matter of principle and argued that 'the best way to finish with the Indians, either by wiping them out completely or driving them to the other side of the Río Negro, is that of offensive war, such as the one waged by Rosas, which nearly finished them off'.[7]

In a private note to himself, Roca described Alsina's ditch as a 'piece of nonsense' worthy only of a 'weak and childish people'.[8] Roca, like many of Alsina's critics, misunderstood the strategic intentions behind his *zanja*. Alsina recognized that Indian raiders could easily cross this barrier, but his ditch was intended to slow them down on their return so that they could be more easily pursued by mobile cavalry detachments. By removing the economic incentive for these raids, Alsina hoped to cut off the trans-Andean 'Chilean road' while simultaneously pushing the frontier southwards. The *zanja* quickly acquired its own symbolic connotations. As Alfredo

Ebelot, the French engineer appointed by Alsina to oversee the construction of the frontier line observed, 'the ditch acquires an almost dramatic interest if one thinks of it as the almost visible limit between civilization and barbarism'.[9]

Calfucurá's successors immediately understood Alsina's proposals. In February 1875, Calfucurá's eldest son Manuel Namuncurá informed the commander of the Bahía Blanca fortress of 'a not very agreeable commission from the Ranquels telling me that the Superior Government was preparing to make war on me' and warned the government that 'if it makes war on me I will know how to defend myself ... I still have enough Indians to defend my lands to the death.'[10] Throughout that year, Namuncurá wrote letters to the archbishop of Buenos Aires, clergymen and frontier commanders expressing his desire for peace while also carrying out raids across the south of the province to reinforce these demands. On 9 May, Bernardo Arriaga, the justice of the peace at the settlement of Tres Arroyos, informed the governor of Buenos Aires Province of three separate Indian 'invasions' inside the 'line of frontier', which had forced the 'great majority' of settlers to abandon their homes. Arriaga concluded his report with a grim warning:

> It is painful but necessary to say it, Minister: the depopulation of the administrative areas of Júarez and Tres Arroyos is a fact, which is unfolding day by day, as a result of the lack of protection pertaining to the interests of the frontier. The desert advances rapidly across the lines that civilisation abandons to it, and whose conquest of the barbarians has cost the country so many sacrifices in blood and treasure. If this state of affairs continues, it is no exaggeration to predict that within a short time these areas will have vanished from the geographical map of the province, and the dominion of civilisation will give way to the Indians of the Pampa.[11]

Today, Tres Arroyos is a prosperous city of 60,000, with giant granaries looming over streets that seem to have been measured to the last millimetre, lined by manicured hedges, where a mock-windmill and clogs outside the railway station pay tribute to the Dutch and Danish immigrants who settled here in the late nineteenth century. In 1875, the population of Tres Arroyos could be counted in the low hundreds, scattered between three small forts, a few houses and the inevitable *pulpería*, encircled by a ditch as protection against Indian raiders. Argentinian politicians often accused Chile of collusion and complicity in these raids, and these suspicions intensified Argentina's determination to impose its own military solutions. On 4 and 5 November 1875, the Argentinian Congress voted in favour of Alsina's ditch.

On 26 December, between 5,000 and 6,000 Indians attacked settlements and forts across a 4,000-mile radius in Buenos Aires Province. Pincén, Baigorria, the Namuncurá brothers and other prominent caciques all took part in a *malón grande* (big raid) that included Mapuche bands from Chile and even 'friendly Indians' from Azul who had rebelled against the government. The frontier corps responded with a counter-offensive in which hundreds of Indians were killed in successive engagements. On 25 April, two army divisions entered Carhué, where their commanding officer Colonel Nicolás Levalle raised the Argentinian flag to mark the permanent conquest for 'civilization and our frontier' of the territory that Calfucurá had defended for so many years. While Levalle's soldiers endured a freezing winter that brought them close to starvation, 4,000 soldiers and labourers began to construct Alsina's barrier.

For two years, soldiers and engineers toiled in the heat and cold while army units relentlessly pursued the caciques who had defied the authority of the government. In October 1877, the Ranquel chief Vicente Pincén humiliated Colonel Conrado 'the Bull' Villegas, the commander of the 3rd Cavalry Regiment, by stealing eighty of the general's prized white horses from their corral at the fort-city of Trenque Lauquen without being detected. Villegas sent sixty-five soldiers to recover the horses, who found them in a *tolderia* some 130 miles away. The soldiers overran the camp, capturing the horses and prisoners and fighting off two separate attacks by Pincén's Ranquels before arriving in Trenque Lauquen in triumph. In 1878, Villegas caught Pincén and brought him back to Buenos Aires, where the newspaper *El Nacional* hailed the capture of 'the most audacious, most reckless, most brave and stubborn Indian'. The satirical magazine *El Mosquito* mockingly described the Ranquel chief's arrival in very different terms:

I have seen Pincén. What a disappointment! He is an old Indian with a long expressionless face, a stupid countenance, childlike eyes that look as if they have been made with a drill, a crushed demeanour, with a jetblack mane of hair despite his age, mute, indifferent to what is happening around him, but with the dignified expression of the savage chief who looks contemptuously at his tormentors and smiles indifferently in the midst of torture, serious, in the way that cretins are with inanimate and insensitive expressions. The Indians that surround him are all as repugnant as he is, with the exception of one captain who at first everyone mistook for Pincén ... As for the women and children ... Ugh! ... let's not speak about that before lunch. Filthy! ... Absolutely disgusting![12]

Pincén's prestige and notoriety were such that Moreno persuaded him to pose for the photographer Antonio Pozzo. The aging cacique was brought to Pozzo's studio and made to stand bare-chested, holding a lance, with a bola hanging round his neck in an attempt to live up to his fearsome image as the 'most-wanted' cacique of the Pampas. The capture of one of the most infamous 'savage chiefs' was another indication that the struggle in the south was turning in the army's favour. Though some *maloneros* crossed Alsina's ditch by forming 'bridges' from piles of dead cattle for their stolen herds to cross, they were hunted down by rapid-response cavalry units armed with Remington rifles that made it difficult for their opponents to get close enough to use their lances and bolas. Telegraph communication between forts increased the speed of these operations. Slowly, but relentlessly, the tide of war was turning in the government's favour, and the limitless open spaces of the Pampas that had so often frustrated Alsina's predecessors no longer provided sanctuary and protection to its Indigenous inhabitants.

* * *

Faced with these tactics, with mounting losses and the near-collapse of the rations system, some caciques expressed nostalgia for the Rosas years. '"Ah! If Don Juan Manuel was alive!"; we have heard the expression in all the toldos', claimed Ebelot in 1875.[13] 'There was never a more sincere desire.' By this time, Rosas had ceased to have anything to do with his native land, as he slipped deeper into poverty at his Swaythling farm to the point that he was forced to sell his clothes, furniture and animals. 'I am sorry to have to tell you that the cows are no longer on this farm', he wrote to his daughter Manuela in London in 1876:

> God only knows that which he disposes: and the pleasure that I felt on seeing them in the field, calling to me, going to my carriage to receive some affectionate ration from my hands, and on sending you all the butter. I've sold them for twenty-seven pounds and if I had waited any longer, they would have offered less.[14]

On 14 March that year, Rosas died from inflammation of the lungs at the age of eighty-three, tended by his doctor, his beloved daughter and his housekeeper. The passing of the dictator was respectfully noted in the local papers. A lone anonymous letter to the *Southampton Times* criticized the attempts to 'whitewash the memory of one of the most cruel, remorseless and sanguinary tyrants that ever existed on earth'. In his last will and testament, Rosas asked to be buried in the Catholic cemetery in Southampton, 'until in my Fatherland, the government recognizes and remembers the justice that my services warrant'. The Argentinian president Avellaneda was not inclined to grant this request to the man who had once had his father killed and decapitated, and Rosas was buried in the Hill Lane Cemetery in Southampton. It was not until 1989 that the Argentinian president Carlos Menem allowed his remains to be returned to Argentina in a gesture of national reconciliation. The dictator's bones were greeted by mounted lancers wearing the uniform of his former militia and laid in the Rosas family tomb in the Recoleta cemetery. But even after his death, the former dictator was not without influence. On 29 December 1877, Alsina died in Carhué from a bladder infection contracted during his many visits to the frontier. Within a week, his place was taken by the man who had criticized his 'wall' and who was now preparing the final destruction of the Pampean Indians with a campaign modelled on the operations Darwin had witnessed in 1833.

PART III

CONQUEST

The frontiers question is the first question for everybody, and we speak about it constantly even if we don't name it. It is the beginning and the end, the alpha and omega ... To subdue the Indians and the frontiers implies nothing less than to populate the desert. We will not eliminate the Indian, without eliminating the desert that engenders him. The fruit will not be extirpated without extirpating the root of the tree that produces it ... We are few and we need to be many. We suffer the evil of the desert and we must learn to conquer it.

President Nicolás Avellaneda, letter to
Colonel Álvaro Barros, 1875[1]

14

THE CONQUEST OF THE DESERT

Until recently, most Argentinians remembered General Julio Roca as the 'Soldier and Statesman, Director of the Campaign of the Desert (1879)' celebrated in the famous image by Juan Manuel Blanes that appears on the old 100-peso notes—showing the general and two-term president on horseback surrounded by his officers at Choele Choel. Nowadays, Roca tends to be treated less reverentially. In the main square of the popular Patagonian tourist resort of Bariloche stands a bronze equestrian statue of Roca in his soldier's cap and coat—the image of dogged tenacity and patriotic resilience in the service of the Fatherland. On any given day, the statue will be splattered with red paint and slogans in Spanish and Mapudungun denouncing Roca as a genocidal murderer and a fascist.[1] In these moments, Roca looks more like a defeated general than a conqueror, hemmed in by dozens of paving stones bearing engraved images of the headscarves of the Mothers of the Plaza de Mayo and the names of the 'disappeared', victims kidnapped and murdered by the 1976–82 junta—a symbolic convergence of David Viñas's description of Argentina's Indians as the country's 'first disappeared'.

This juxtaposition is not coincidental. In Argentina, the Czech novelist Milan Kundera's formulation of the struggle against political power as a struggle between memory and forgetting is waged more fiercely than most. Murals and monuments commemorating victims of the dictatorship can be found all over the country. These campaigns of remembrance also extend to the nineteenth-century campaigns against Argentina's Indigenous Peoples in calls to change the names of streets and public places named after generals and politicians associated with these campaigns, and such efforts have often focused on Roca. In 1996, the late historian-activist Osvaldo Bayer called for a statue of Roca to be taken down from a Buenos Aires street, and in the years that followed various towns and cities renamed streets named after the hero-general. The writer and activist Marcelo Valko accompanied Bayer during many of these campaigns; one wall of his living room in Buenos Aires is decorated with signs taken down from different parts of Argentina. We discussed these campaigns and his latest book—a tenaciously researched and demythologizing account of the Conquest of the Desert.[2] 'The thesis of my book is very simple,' he says. 'There are many people on pedestals who deserve a criminal record. They shouldn't be up there.'

For Valko, the demonumentalization of Roca is both an act of historic reparation for Argentina's First Peoples and a rejection of what he calls *desmemoria*—a word that means 'forgetfulness' in the sense of 'deliberately forgetting'—in a country that treats its population 'like dogs'. He sees Argentina's 'statue wars' as part of a global phenomenon that includes campaigns against Confederate flags and war heroes and Edward Colston's statue in Bristol. 'It doesn't change history,' he says. 'The slaveowner exists, but he can't be honoured. What we say is "justice can't be achieved for those who have died, but we can punish the memory of their persecutors".'

Such arguments have not gone unchallenged. In an article in *La Nación* in November 2004, Juan José Cresto, the-then director of Argentina's National History Museum, described the accusations of genocide directed at Roca as 'historiography lacking any documentation' intended to promote what he regarded as illegitimate Mapuche land claims—a claim rejected in an open letter written by more than a dozen historians.[3] In 2021, the journalist and radio presenter Rolando Hanglin, vice president of the Instituto de Estudios Históricos Julio Argentino Roca (Institute of Historical Studies Julio Argentino Roca), described the attacks on Roca as a 'Mapuche lie'. As is often the case elsewhere, these conflicting versions of the national past tend to break down on left/ right political lines, but in Argentina it is only relatively recently that these debates have touched on what was once regarded as a glorious episode in the history of the nation.

* * *

The object of this veneration and opprobrium was born into a military family from the north-western sugar-growing province of Túcuman in 1843. At the age of thirteen, Roca was admitted to the prestigious Colegio del Uruguay as a military cadet. By the time he was seventeen, Roca had served as a junior artillery officer at the Battles of Cepeda and Pavón, and he went on to serve in the Paraguayan War. Despite his admiration of Rosas's Desert Campaign, Roca was a very different character. The conservative nationalist and political biographer Mario Octavio Amadeo (1911–83) commented on Roca's 'mimetism, his adaptability', his 'great topographical memory' and his 'moderate and sober' temperament.[4] Others have described the soft-spoken officer as a 'gentleman farmer of exquisite manners' with a fondness for chess, Molière and Napoleon; as cynical, devious and politically ambitious—a reputation that earned him the nickname 'el Zorrito'—'the Little Fox'.

Following Alsina's death in 1877, Roca became minister of war at the age of thirty-four, and he proceeded to make his case for the general offensive against the Pampean caciques that he had outlined to his predecessor. In his first address to Congress, he rejected 'the old system of successive stages of occupation handed from the Conquest, obliging us to scatter our national military forces over a very broad area and to remain vulnerable to all Indian attacks'. Instead, Roca proposed to 'seek the Indian in his encampment, to defeat him or drive him out, leaving him barred not only by a large man-made ditch but by the great and impassable barrier of the Río Negro'.[5] Beginning in July 1878, the Argentine army launched a series of expeditions throughout the year that broke the resistance of the Mapuche and Ranquel peoples. Though Roca promised to respect the treaties signed with loyal Pampean caciques, he warned the Ranquel chief Manuel Baigorria that 'unfriendly Indians' would be treated 'without mercy, even to their extermination'. These distinctions were not always observed. In November that year, an army unit under the command of Roca's brother Rudecindo massacred some sixty Ranquel men who had come peacefully to claim rations at the city of Villa Mercedes. The massacre was widely condemned by the Buenos Aires press, but neither Roca's brother nor anyone else suffered any negative consequences for it.

In 1878, Zeballos was invited by Roca to compile the historical and scientific data on the 'empire of the Pampa' to demonstrate the viability of Roca's military solution to the 'frontiers question' to Congress. Drawing heavily on the writings of FitzRoy, Musters, Mansilla and other travellers, Zeballos travelled across the Pampas, compiling navigational charts and topographical and meteorological data. These journeys resulted in his bestselling *La conquista de quince mil leguas* (The conquest of 15,000 leagues, 1878), in which he crowed: 'The military power of the barbarians is completely destroyed, because the Remington has taught them that a battalion of the Republic can cross the entire Pampas, leaving the country sown with the corpses of those who dared oppose it.'

Bolstered by this assessment, Roca obtained a one-off payment of 1.6 million pesos from Congress to finance the 'Campaña Expedicionaria al Desierto' (Expeditionary Campaign to the Desert). In January 1879, the Ministry of War reported that 5,077 Indians had been captured or killed in the previous year, compared with only thirteen government casualties. In April that year, five divisions marched southwards in separate columns from different points along the length of the frontier, just as Rosas's armies had done before them. On 26 April, Roca told his soldiers:

> The United States is one of the most powerful countries in the world, and yet until now it has not been able to solve the Indian problem, even though it has tried all kinds of strategies, has spent millions of dollars annually, and has dispatched numerous armies. You, the Argentine army, are now going to solve that problem at the other end of America, with a small expenditure of your courage.[67]

On 29 April, 6,000 soldiers, including hundreds of 'friendly Indians', marched south, accompanied by priests, scientists, savants, journalists, a photographer and expedition family members in what Roca called 'the exodus of a people in motion' into the Pampas. Roca accompanied Villegas's 1st Division, heading south from Azul along the route taken by Rosas in 1833. Like Rosas before him, Roca saw the conquest of the south as a means of furthering his presidential ambitions, and these objectives were made clear as his columns converged on the 'bandit capital' of Indigenous Patagonia that Rosas's armies had taken in 1833 and subsequently abandoned.

* * *

The town of Choele Choel is located 192 miles south-west of Bahía Blanca on the edge of the Río Negro, alongside the Route 22 that connects the Atlantic coast to the province of Neuquén. Walk a few blocks southwards from the main highway through the tree-lined streets, and you reach the bridge that crosses the narrower part of the Río Negro and leads to the island of Choele Choel in the middle of the river. Today, the island is a recreation and picnic area, taken up by sports fields and a camping ground where rows of picnic tables and barbecues stand in the shade of weeping willows. Clouds of mosquitoes follow the unlucky visitor without protection, and swimmers need to be wary of the fastmoving currents that bear the force of two rivers, the Limay and Neuquén, as the Río Negro surges eastwards towards the Atlantic. It is difficult to imagine the dread and horror with which white Argentina once regarded this little island. For much of the nineteenth century, Choele Choel played a central role in Argentina's Indian wars. Inter-tribal parliaments were held here, and the island was also used as a natural corral, where captives, cattle and horses were rounded up before being taken across the 'Chilean road' across the Andes.

There is no record of this history on the island or in the town itself, but about a mile and a half westwards an imposing ziggurat stands on a low plateau overlooking Route 22, with a red stain running down its length. You have to walk up the hill to see the graffiti slogans denouncing Roca as a genocidaire, the streak of red paint and the barely legible inscription: 'The Argentine people to General Julio A. Roca, to the Army and Navy Expeditionaries, who incorporated Patagonia into the activity of the Nation. 30th of November 1878–24th May 1879.' On the front façade, the Argentine national shield faces southwards towards the Patagonian steppe. On the eastern side, caryatids of two brawny cloth-camped workers in the brutish socialist realist style stare in the same direction, holding a sextant and a plough. A tree and a cow complete the symbolism—the productive forces unleashed by the settlers and immigrants who wrested the desert from barbarism. On the opposite side, two Indians stand with their horses, holding a spear and a bow—a weapon that was not used in the Pampas. This overwrought allegory commemorates the arrival of Roca's soldiers at this same point on 24 May 1879. Climb the steep winding steps to the top of the tower, and Patagonia stretches out beyond the fertile green banks of the Río Negro across the far horizon, much as it once would have looked to Roca's conquering army.

Roca went to great lengths to reach the island in time for 25 May—the date of the 1810 'May Revolution' against Spanish rule—and his troops' arrival was immediately transmitted by telegraph to Buenos Aires, where it received an ecstatic response. President Avellaneda congratulated Roca by telegram on 'the superb outcome of your enterprise, for the exactitude of your operations, for the perfection of all the military services, for the indefatigable constancy of their soldier, and for the skill of the leaders, never so well demonstrated as on this occasion'. The newspaper *El Siglo* hailed 'the vast plan conceived by General Roca' as 'a game of chess played on the vast board of the Pampas. Civilization has won splendidly. It has won for our progress, for our future.' At an official reception in Potsdam, the Prussian general Helmuth von Moltke congratulated the Argentinian plenipotentiary Miguel Cané on the brilliance of Roca's campaign.

Few questioned the triumphalism of a campaign that achieved its objectives without a single battle. No sooner had Roca's soldiers begun to establish a base on the island than the Little Fox was whisked off by steamship to Carmen de Patagones, and then to Buenos Aires, to accept the accolades of a grateful city and begin his presidential campaign. In July that year, the Río Negro began to rise, lapping against the earthen barrier that Roca's army had hastily constructed. For two weeks, military bands played music and soldiers performed drill exercises in the mud and sang the national anthem in an attempt to raise the morale of what the garrison's commanding officer called 'an army division ... on the point of disappearing ... not vanquished by men but encircled and exterminated by God'. In the event, the floodwaters receded without breaching the parapet, and the army established a permanent presence in the bandit capital of the Pampas that is now given over to football and barbecues.

* * *

This near-flooding was the only significant threat any of Roca's columns faced in their march to the south. Worn down by years of war, weakened by hunger and an epidemic of smallpox, the scattered remnants of Calfucurá's once-dominant confederation either took flight or surrendered to Roca's armies. On 15 July, a unit from the 4th Division trapped the Ranquel chieftain Manuel Baigorria and his family near the banks of the Neuquén. Baigorria managed to escape with some of his captains and warriors, but a cavalry unit caught up with him shortly afterwards. With his horse shot from under him, the chief known as the 'white cacique' because of his creole origins faced his pursuers on foot with his lance in hand and died in a hail of bullets. The official campaign report of 1879 lists 1,313 warriors killed and another 1,271 captured among the 14,000 prisoners that consisted mostly of women, children and the elderly known as the *chusma* (rabble).[8] There were only fourteen casualties on the government side.

By the end of that year, army units had already begun to carry out 'reconnaissance and occupation' operations south of the Río Negro in accordance with Avellaneda's orders that 'not a single Indian tribe will remain that has not been subjugated throughout the Pampas and Patagonia'. In April 1880, Roca won the presidential elections, and the following year an army unit commanded by Villegas reached the shores of Lake Nahuel Huapi in northern Patagonia. Villegas's campaign journal combined accounts of sporadic military confrontations with detailed descriptions of Patagonia's topography, minerology and the region's 'immense riches, concealed by their savage inhabitants, but which will not pass unnoticed by the investigative and studious gaze of civilization'.[9] Between 1883 and 1884, the army extended its operations to the Mapuche-Tehuelche 'Land of Apples' in northern Chubut in pursuit of the caciques Valentín Sayhuque, Modesto Inacayal and Foyel Payllakamino. These campaigns demonstrated how far Roca had departed from

Rosas's carrot-and-stick model. Inacayal had facilitated Moreno's explorations of northern Patagonia with Sarmiento at the Moreno family home. Sayhueque had resisted repeated overtures from Calfucurá and his successors to join forces against the *huinca*, and he had also rejected requests from Chilean government envoys to fly the Chilean flag from his camp on the grounds that 'he was Argentinian and therefore only raised the flag of his country'.

In a letter to his friend Lewis Jones, the leader of the Welsh colony in Chubut, Sayhueque bitterly complained: 'I have not broken the peace and goodwill that exists between myself and the Government for more than twenty years, and I have faithfully fulfilled all of the agreements entered into at Patagones.' Sayhueque railed against the soldiers who 'came armed stealing into our living tents, as if I was an enemy and a murderer', even though he was 'born and raised on the land and an Argentinian faithful to the government', and he pleaded with Jones to intercede with the authorities on his behalf.[10] It is not known whether Jones acceded to this request, but for Roca and his commanders, the 'humiliation' of *negocio pacífico* with Indians was now a thing of the past. The invasion of northern Patagonia saw some of the bloodiest engagements of the conquest, as Sayhueque and other chiefs used the densely wooded mountains of northern Chubut as a base for guerrilla warfare against their pursuers.

Between December 1882 and April 1883, 364 Indians were killed in combat in the Andean foothills, in addition to an estimated 3,000 people who 'disappeared from the captured terrain', according to Villegas. On 9 February 1883, an army unit commanded by Lieutenant-Colonel Nicolás Palacios attacked a 1,000-strong Indian camp at the Río Apeleg, near Lake Nahuel Huapi. A three-hour gunfight ensued in which Indian women fought Palacio's mounted soldiers with shearing scissors. Though Inacayal escaped, between eighty to 100 members of his camp were killed. Over the next two years, the army continued to carry out 'raid, exploration and reconnaissance' operations in the mountains and lakes between the Senguer and the Chubut rivers and Puerto Deseado on the Atlantic Ocean.

In November 1883, Namuncurá formally surrendered to General Lorenzo Vintter, the governor of Patagonia. In 1884, Inacayal and Foyel also surrendered, and on 1 January 1885 Sayhueque gave himself up at Junín de los Andes, along with 700 warriors and 2,500 *chusma*. In a report to the minister of war that year, Vintter informed his superiors that 'every borderline limitation against the savage had disappeared forever from the South of the Republic'. Vintter praised the 'active operations against the savage' begun by Alsina in 1876 and continued under his successor, which had placed a vast territory of more than 40,000 square leagues under the control of the Argentinian state. The following year, the outgoing president Roca gave his last address to Congress, in which he hailed the subjugation of the Indigenous territories in Patagonia and the south and in the Chaco region of north-eastern Argentina. All these territories, Roca declared, had been incorporated into the 'dominion of civilization' and their 'forests, their rivers, their most extensive grasslands, their minerals' incorporated into new provinces as a result of their 'emancipation from the power of the savage'.[11]

* * *

Roca's peers were equally exultant at what Colonel Manuel Olascoaga, an officer in the 1st Division, called 'the pacification of the desert regions to the south of the Republic; the conquest of twenty thousand leagues of fertile land handed over to civilization; the submission and regeneration of savage populations; the liberation of several hundred captives; the conclusion of the century-old war against the Indians'.[12] *El Siglo* celebrated Patagonia's emergence from 'obscurity and oblivion', in which 'the solitudes of the desert gave way to the thunderous noise of civilization'. The cost of Roca's campaigns was underwritten by auctions and bonds worth a square league (more than 7,000 acres), whose holders chose lots that had already been surveyed even before Roca's armies reached them. The Sociedad Rural raised bonds based on land sales of 4 pesos per hectare, while advertisements in Buenos Aires newspapers offered the opportunity to acquire 'public lands' for 10,000 pesos a square league.

'Never has a safer and more brilliant business opportunity presented itself to landowners, capitalists and landlords', declared *El Siglo*, and these were the sectors that benefitted most from the conquest. In an 1882 land auction, holdings were limited to 100,000 acres per person, but speculators used agents and false names to circumvent the rules. By 1890, according to the historian James Scobie, 'the whole area of the pampas had passed into private hands, to be held for speculation, investment, or prestige, but not to be owned by those who cultivated the land'.[13] Most of the 54 million hectares that became part of Argentina's 'national patrimony' in 1878–9 were bought by Argentina's richest families or by foreign—particularly English—landowners, to be turned into cattle ranches or sheep stations. In theory, every soldier who participated in the campaign was supposed to receive a grant of land, but officers tended to buy these lots at auctions before selling them on to speculators. While officers founded new towns in the conquered territories or became provincial governors, most ordinary soldiers ended up as poor at the end of these campaigns as they were when they began.

For the Mapuche, Tehuelche, Ranquels and other Indigenous Peoples, the Conquest of the Desert brought death, hunger and enslavement. At least 10,313 Indigenous men, women and children were force-marched to Valcheta, Carmen de Patagones and Bahía Blanca, where they were shipped to Buenos Aires or taken to military barracks, prisons and other improvised *depositos de indios* (deposits of Indians) across the country. The Welsh settler John Daniel Evans describes a visit to an Indian reformatory/prison near Valcheta, where Indian prisoners behind a high barbed-wire fence begged him and his companions for bread. Among these prisoners, Evans recognized a 'childhood friend, my brother of the desert with whom we had always shared what bread we had'. Evans had nothing to alleviate his hunger and resolved to return from Valcheta with money in the hope of bribing the guards to allow his friend to leave, but his friend died before he was able to save him.[14] These 'deposits' appear to have been part of a pre-planned policy of removing the entire Indigenous population from the lands they occupied.

Many of these prisoners and deportees were already weakened by hunger and smallpox before they were captured. Between January and November 1879, more than 800 Indians were baptized on Martín García Island, in the Río de la Plata, most of whom were declared *in articulo mortis* (at the point of death) when these baptisms occurred. In a letter to his superiors in October 1878, the island's only doctor warned that 148 Indian prisoners had arrived on the island infected with smallpox who were too malnourished to work, and that in addition to 'the sadness that they carry, the moral despondency, they feel the loss of the desert'.[15] These outcomes may not have been intentional, but Roca's officers did little to prevent them. Unlike the US government, Argentina's rulers did not—with few exceptions—create self-governing Indian reservations. In effect, the army acted as the blunt instrument of a policy of forced assimilation.

Work—whether paid or unpaid—was seen as the key to the civilizing process. Indigenous prisoners at Martín García Island broke rocks into cobblestones for the Buenos Aires streets. Others were forced into plantation labour on the sugar refineries of Túcuman Province or press-ganged into the army or the navy, where they were given new names such as Bismarck, Gambetta or Garibaldi to signify their transformation into *indios marineros* (Indian sailors). Buenos Aires newspapers advertised weekly *repartos* (auctions or distributions) of Indians on the docks of the River Plate, organized by the Sociedad de Beneficencia (Society of Beneficence), where wealthy families selected servants from the women and children who arrived at the docks. For Olascoaga, these rituals constituted 'the most satisfying and moralizing spectacle that could offer itself to civilized people; the patient transformation of barbarism into civilization; the visible moment of dignifying humanity; the palpable deed of converting the element of destruction into an element of progress'.[16]

* * *

Not all of Olascoaga's contemporaries shared this assessment. One horrified Buenos Aires reporter described 'the desperation, the crying [that] does not stop, children are taken away from their mothers because they are given away as presents in their presence, despite the cries, the screams and the begging that, with their arms aiming at the sky, these Indian women shout'. In 1884, Senator Aristóbulo del Valle told Congress:

> We have taken families from the savages, we have brought them to this center of civilization, where every right seems to be guaranteed, and yet we have not respected for these families any of

the rights that belong, not only to civilized men, but to humanity: we have enslaved the men, prostituted the women, we have torn the children away from their mothers, we have sent old men to work as slaves anywhere. In a word, we have turned our backs and broken all the laws that govern the moral actions of men.[17]

Such criticisms failed to deflate the euphoric celebration of Roca's 'game of chess' as the culminating victory of civilization over barbarism. Until recently, this was the version of Roca's campaigns that passed into posterity. The 1976–82 military dictatorship frequently drew comparisons between Roca's campaigns against 'savages' and its own 'dirty war' against left-wing 'subversives'. The official guide for the 1978 World Cup hailed Argentina's success in 'conquering the lands of indigenous people to incorporate them into the nation'. During the centenary of the conquest in 1979, the dictatorship staged a series of commemorative events, including a military parade in Choele Choel to mark Roca's arrival on the island. Leading members of the military compared the 'foreign' Indians of the Pampas to the left-wing 'subversives' of their own era.

The vanquished have very different memories of these campaigns. In 1899, a document written in Mapudungun was sent to the Chilean linguist Rodolfo Lenz that described an 'imaginary parliament' composed of mostly deceased caciques from the Puelmapu—the eastern (Argentinian) territories of the larger Mapuche entity of Wallmapu. Written as a polyphonic narrative by a Mapuche named Ignacio Cañumir, each chief recounts events that took place in the years leading up to the Conquest of the Desert and beyond. Their voices tell a collective story of war, defeat, raids and invasion; of the loss of their animals, their wives and their land; of forced marches, imprisonment and the separation of families.[18] The *ülkantun* traditional songs referring to the conquest strike a similarly threnodic note. 'That land has been lost / The end of that land has arrived / Now the invaders come', laments one song: 'You must have strength and let's run / The invaders are coming / The invaders are coming.'

These oral histories have only gradually filtered into the wider memory of the Conquest of the Desert. In 1906, the German anthropologist Robert Lehmann-Nitsche (1872–1938) interviewed survivors of the conquest, which were not published until 2013 when two Chilean-Mapuche researchers discovered them among Lehmann-Nitsche's papers in Berlin. One of Lehmann-Nitsche's interviewees was a Mapuche man named Katrülaf, who describes his capture along with Inacayal and Foyel in 1884, near one of the Welsh settlements in Chubut Province. Katrülaf tenderly depicts the life that he and his people lived before the conquest, with its freedom and abundant game, before recounting his brutal odyssey under military escort through Chubut Province to Valcheta, Patagones and Buenos Aires, along with the caciques Inacayal, Foyel and their families. Chained up for weeks on end, and subject to forced marches without food and water, Katrülaf and his companions were finally taken to Buenos Aires and imprisoned in a military barracks.[19]

Katrülaf was eventually allowed to return to his lands after six years in captivity. Other captured Indians had very different fates. Pincén was imprisoned on Martín García Island, and various unverified reports claim that he was killed by the army. Others claim he managed to escape and returned to the Pampas. In 1884, Calfucurá's son, Namuncurá, was given the title 'Colonel of the Nation' and allowed to return to live out the rest of his days on a government pension. His son Ceferino became a Catholic priest, and in 2007 the 'Lily of Patagonia' was beatified by Pope Benedict XVI. Epumer Rosas or Epu Ngürü ('Two Foxes'), the brother of Rosas's godson Mariano, was released from Martín García Island in 1882 and worked on a ranch, where he died of hepatitis two years later. One of his grandsons, Juan Epumer, went on to become a guitarist and accompanist of the great tango singer Agustín Magaldi. This began a musical dynasty that includes the jazz guitarist Lito Epumer and the late María Gabriela Epumer—a singer and guitarist with the famed Argentinian singer-songwriter Charly García. Sayhueque remained for nearly four months in Buenos Aires before he was allowed to return to an arid corner of Chubut Province. In 1903, he died on his *tolderia* on the banks of the Río Genoa. In a telegram to President Roca, the Salesian priest Lino Carbajal claimed that Sayhueque had died as a Christian, and that 'with him dies the last vestige of that savage empire that the valiant army conquers for civilization on the orders of Your Excellency'.[20]

In the aftermath of Roca's campaigns, all the Indigenous Peoples of the Pampas and Patagonia were forced to build new lives in the society of their conquerors. Some returned to a small part of the lands they had once occupied, policed by military forts. Others migrated to the cities and became part of the urbanized rural population for whom middle- and upper-class Argentines in the 1940s coined the disparaging label *cabecitas negras* (little black heads)—a catch-all term incorporating 'Indian', 'black' and 'poor'. It was not until Argentina's democratic transition that the descendants of these campaigns began to emerge from obscurity and make their voices heard.

I met Fabiana Méndez Calfunao ('Blue Tiger') in a café on the corner of the Plaza Rivadavia in Bahía Blanca. A vibrant and thoughtful Mapuche woman who looks younger than her fifty-four years, Fabiana described how her mother was taken into domestic servitude in 1939 at the age of twelve or thirteen by a landowner who later married her. Her mother had three children before she managed to escape with the help of her grandmother. 'What happened to my mother happened to many *abuelas* [grandmothers] I know,' she said. 'This splitting up of families, this separation of children, taking them to concentration camps. And of course if they separate you from your family, they annihilate you.' This history was largely absent from Fabiana's primary schooldays, where she was taught

> by a nun using a manual that showed pictures of Indigenous People with the cape and feathers, and the lance ready to kill—always savage. They taught me at primary school that the Indian had to be killed, that it was necessary to enable economic expansion and industrial production, because the Indian was a savage who wouldn't allow progress.

Fabiana's father moved to Bahía Blanca in 1958 to work on the railway. Fabiana remembers a painful childhood, where her classmates treated her as 'this lice-ridden, smelly, bad Indian'. Now she speaks with pride about her Mapuche origins. In addition to her work as a seamstress, she also gives workshops on ceramics and silverwork in Mapudungun to young people and volunteers for a local organization that promotes Mapuche culture. She was cautiously optimistic about the future of the Mapuche but described the Conquest of the Desert unequivocally as a form of 'social destruction' and as a 'terracide'—the murder of the earth. 'Everything that was done was hidden,' she said. 'Today, words like decolonization, interculturality and genocide are words that reflect evolution and struggle and can be spoken.'

Not all Argentinians accept these developments. In 2021, a Molotov cocktail was thrown at the house of a local Mapuche community activist in Bahía Blanca, and a group calling itself the Comando de Restauración Nacional (National Restoration Command) left a pamphlet declaring war on the Mapuche and other groups who had 'humiliated and impoverished' the 'great hero-patriots of Argentinian history'. That same year, the Bahía Blanca town council held an internet poll to give the city's 'Conquest of the Desert' park a name that did not refer to the 'violent acts' committed against First Peoples. This outcome was not successful. The winning name—chosen by internet poll—was 'Julio Argentino Roca'. Such polls are always easy to manipulate, but the result was another sign that Argentina's history remains a contested subject in which the past and present still collide.

15

THE TEMPLE OF EVOLUTION

Darwin never commented publicly on the Conquest of the Desert. When Roca's armies reached the Río Negro, Darwin was seventy years old and in poor health, but he still took a lively interest in world affairs, and Argentina and Patagonia were still very much on his mind. In 1876, he wrote a short autobiography for his children, in which he described once again 'the sense of sublimity, which the great deserts of Patagonia and the forest-clad mountains of Tierra del Fuego excited in me' and recalled 'the sight of a naked savage in his native land' as 'an event which can never be forgotten'.[1] In 1878, he was awarded an honorary diploma by the Argentinian National Academy of Sciences, and in March the following year he wrote a letter of thanks to Hendrik Weyenbergh, the Dutch-born zoologist and president of the academy, informing him that he had instructed his publisher to send a copy of *Origin* to the academy 'as I suppose that this is the best of my works'. In 1880, he exchanged letters with the Scottish writer Lady Dixie, who had recently followed his footsteps in Patagonia. Darwin also corresponded regularly about the Fuegians with his former shipmate Sulivan. And yet there is nothing in the historical record to indicate he was aware of the military campaigns that were so closely modelled on the 'war against barbarians' that he had observed in 1833.

If Darwin was unaware of these campaigns, his Argentinian followers were very much aware of his connection to their country. In an obituary following Darwin's death on 19 April 1882, Sarmiento described Argentina as 'the first theater of his works' and hailed the conquest of Patagonia as the completion of the process of discovery that Darwin had begun. 'Today this terra incognita is the southern limit of Argentine territory,' Sarmiento wrote,

> and now that its great explorer has departed the last beach of life to penetrate into the unknown reaches of death, his shade remains to guide the future settlers of that country, where in his honor, a mountain bears his glorious name ... How satisfying it is for a new nation such as our own to find its own name attached to the golden link of Darwin's, and to the fruitful labors of that young Darwin of half a century ago![2]

This 'golden link' was not only based on Darwin's personal connection to Patagonia. When Roca's armies marched south, they were accompanied by a Scientific Commission of botanists, zoologists, ornithologists, astronomers, geologists and geographers. In its final report on the expedition to the Argentine Senate, the commission described the 14,000 Indians killed or taken prisoner as a testament to the 'intellectual superiority, activity and enlightenment' of the conquerors, who had removed the 'sterile race' that occupied these territories and 'broadened the horizons of the future and brought forth new sources of production for humanity'. The German-Argentine botanist Paul Günther Lorenz praised Roca for having brought 'civilization and industry, as well as science, to these vast areas', while the chemist and zoologist Adolfo Doering hailed the collaboration between scientists and soldiers and expressed 'our fervent hope ... that the resplendence of Mars and the radiance of Minerva shall increasingly form an inextinguishable halo'.[3]

* * *

In July 1881, in a letter to the philosopher William Graham, Darwin took issue with Graham's suggestion that natural selection had not been relevant to the 'progress of civilisation'. Darwin reminded Graham once again how 'the more civilised so-called Caucasian races have beaten the Turkish hollow in the struggle for existence. Looking to the world at no very distant date, what an endless number of the lower races will have been eliminated by the higher civilised races throughout the world.'[4] Many of Darwin's Argentinian followers understood the Conquest of the Desert in precisely these terms. In a newspaper article celebrating Roca's campaigns in 1879, Sarmiento predicted that 'the savages will be unable to recover for many years to come' because 'it is a fatal flaw of savage life, that they can never recoup their losses, for once they have come into contact with civilized peoples, they are condemned to final extinction'. For Alsina's French engineer Ebelot, Argentina's 'brutal and ferocious savages' and 'degraded races' were destined to succumb to 'the laws of an ascendant evolution'.[5] In May 1882, the naturalist, novelist and explorer Eduardo Ladislao Holmberg (1852–1937) gave a lecture on Darwin to a packed audience at the Teatro Cólon in which he attributed the spread of the black thistle, which was displacing local varieties in Buenos Aires Province, as an example of natural selection. Holmberg proceeded to give another example with which his audience would have been more familiar:

> In the Animal Kingdom, we have the Indian. Is his cause just? To put the matter bluntly, the Indian is defending his land, which we have usurped, and so he hurts, robs us, and kills us. But does he do right? It's not always clear. This is the struggle for life.[6]

Unlike some of his contemporaries, Holmberg did not depict Argentina's victory in this struggle as a triumph of civilization but as the result of a natural process in which '[w]e whites, we civilized Christians, armed with our Remingtons, shall do away with the Indians, because the law of Malthus stands above all individual opinions'. In this 'victory of organic process', the white population 'armed with good ideas, weapons, and resources' had triumphed over another. 'Do we do right?' Holmberg asked his audience. 'That's a question. "We struggle for life." That's an answer.' Moreno also described the Argentine Republic as 'a vast necropolis of lost races. Stemming from very remote areas, pushed by the fatal struggle for life, in which the strongest prevail, some arrived as victors, and they annihilated [each other] in our extreme south.'[7] If the Conquest of the Desert was the outcome of the laws of evolution, it also made the 'necropolis of lost races' accessible to those who wanted to study them. Before 1879, scientists and explorers were obliged to seek permission from Indigenous caciques to enter their territories. Now the entire south was at the disposal of Argentinian science, and it was in these circumstances that Moreno was able to persuade his government to build a new temple to science, and to Darwinian science in particular.

* * *

It takes just over an hour by bus to reach the city of La Plata, the capital of Buenos Aires Province, from Buenos Aires itself. The city was officially founded in 1882, following the federalization of the city of Buenos Aires that finally resolved the long-running conflicts between Buenos Aires and the provinces that had destabilized Argentina since independence. Photographs taken by the army photographer Tomás Bradley between 1882 and 1885 show a mostly empty city springing out of the flat plain, with no obvious connection to anything around it. Today, La Plata is an administrative city of 772,000, and its main attraction is the Museo de La Plata, a natural science and anthropological museum that forms part of the Faculty of Natural Sciences of the University of La Plata. Its existence is almost entirely due to Moreno's lobbying efforts and his family's political connections. Before the Conquest of the Desert, Moreno had unsuccessfully tried to persuade the government to build a more permanent home for the collection of fossils, Indigenous artefacts and osteological remains that he exhibited on a single floor of the Teatro Colón.

In 1880–1, Moreno visited museums and anthropological institutes in Paris and London, and he returned to Argentina determined to establish a similar museum in his native land. In an address to the Sociedad Científica in 1882, he called for a permanent anthropological museum that would reveal to contemporary 'Argentinian man' the

> long sequence of the physical and social evolution of his predecessors, from that humble primitive animal, physical man, who made no use of the intellectual spark inside his brain, all the way to the great, wise, conquering legislator, who raised the cities now scattered in ruins throughout the territory of the Republic.[8]

At that time, Moreno's Anthropological Museum at the Teatro Colón contained some 15,000 pieces, including 400 prehistoric skulls belonging to 'ancient extinguished races' and another hundred that he categorized as *indigenas actuales* (current Indigenous Peoples).

With the establishment of the provincial capital, the government finally agreed to build a new museum in La Plata. In curating the museum's first collections, Moreno collaborated closely with the palaeontologist Florentino Ameghino. Both Moreno and Ameghino saw the museum as an explicitly Darwinist rival to the Public Museum of Buenos Aires and its conservative director Dr Hermann Burmeister, as well as an opportunity to open a new chapter in Argentinian science. In 1884, Moreno's project was approved, and three years later the museum was completed. Its design incorporated many of the ideas Moreno had gathered in his European travels, and its long rectangular building with its semi-circular galleries on either side and granite neo-classical façade remain one of the most impressive sights in La Plata. Rising up from rows of planted trees and a man-made lake, the mock-Corinthian pillars and wide steps echo the 'chapels of humanity' built by the followers of the positivist philosopher Auguste Comte in Brazil and other countries. But the stone sabre-toothed Smilodons, and the busts of scientists and explorers, including Darwin, Humboldt and FitzRoy, all testify to its original purpose as a monument to evolution and to the 'heroes of science' Moreno admired.

In the rounded reception hall with its glass dome ceiling, Moreno's bust stands facing the doorway, surrounded by romantic timeless frescoes of Indians on horseback carrying lances, hunting guanaco with bolas and holding parliaments in an open plain. From the rotunda, marble corridors lead off into exhibition halls filled with the models and remains of Mylodons and Mylodontidae, Toxodons and giant sloths and armadillos, and giant lizards, hanging from the ceiling or displayed in vitrines. It's a beguiling and immersive spectacle, dramatized by numerous tableaux designed to take the visitor on an imaginative and scientific journey from the earliest life forms to the present. The museum's otherworldliness is enhanced by its old-fashioned wooden panelling, its floor-to-ceiling cabinets, its stained-glass windows, its brightly lit display cabinets and the warm natural light through the tiled glass ceiling.

Nineteenth-century visitors followed the circular corridors around the building, in accordance with Moreno's concept of a 'biological ring' that traced the evolution of species from their most primitive and basic forms and culminated in the anatomical, anthropological and ethnological galleries on the first floor. These galleries housed thousands of pieces from Moreno's personal collection of skulls and osteological remains, augmented by contributions from the network of collectors established and trained by Moreno in Argentina and the Southern Cone countries. In 1889, Zeballos donated sixty-four skulls and hundreds of geological and paleontological pieces to the museum's 'American Anatomy' gallery, which included the remains of Juan Calfucurá.

Many of these exhibits were acquired through excavations of tombs and burial grounds in the newly conquered southern territories. Calfucurá's remains were dug up by the army in his former stronghold in Carhué and given to Zeballos as a present. In 1888, the museum photographer Santiago Pozzi returned from a scientific expedition to the Patagonian province of Santa Cruz with the skull of an Indigenous man named 'Michel' who had accidentally been shot by a member of his expedition. Moreno's secretary, the Italian-Argentine naturalist Clemente Onelli (1864–1924), was hosted by a Tehuelche chief named Klocosk in northern Patagonia. When his host died unexpectedly, Onelli noted the burial site and returned to the spot a year later to 'unbury the skull that has enriched my anthropological collection'.[9] In 1907, the botanist Carlos Spegazzini donated the skull of an Indigenous woman he had retrieved from a Patagonian ravine in 1888, whom he had known personally before she had died from unknown causes.

* * *

For Moreno and his fellow-collectors, death instantly transformed Indigenous individuals into scientific artefacts, regardless of their wishes in the matter. In 1905, the Catalan lawyer, economist and writer Federico Rahola y Tremols (1859–1919) described his shock on seeing 'the men who only yesterday defended their native land against the invader' reduced to 'a mere archaeological curiosity' in which 'the cemeteries that conserved the remains of their ancestors close to their camps, are grouped and classified in showcases, presenting the unusual case of a people sacrificed in the name of civilization, dispossessed of their land, whose remains have since served to form the collections of a zoological museum'.[10] By 1910, the museum's catalogue listed 5,588 skeletons, skulls, pieces of bone, mummified remains, scalps and brains, many of which were displayed in vitrines along with Calfucurá, Cipriano Catriel, Mariano Rosas and other caciques who had only recently constituted enemies of the Argentine state.

The remains of 'friendly Indians' also found their way into the museum. In July 1883, the gentle Tehuelche chief Orkeke, Musters' friend and guide during his journeys through Patagonia, was detained by the army along with his family and fifty-four members of his tribe near Puerto Deseado. Orkeke and his people had never raised a hand against the government and flew two Argentinian flags at their *tolderia* in a demonstration of loyalty, yet their horses were taken from them, and they were shipped to Buenos Aires accompanied by General Vintter and held in the barracks of the 1st Artillery Regiment. Orkeke managed to obtain a meeting with President Roca, who described his imprisonment as a 'mistake' and ordered the army to return him to Patagonia. In the meantime, the Tehuelches were hosted by the soldier and explorer Ramón Lista (1856–97), who took them on visits to the zoo, sweetshops and the theatre.

Orkeke and his entourage became an exotic object of media and public fascination. On one occasion, they were brought on stage after watching a light opera, where they sang Tehuelche songs to a delighted audience that showered them with money and sweets. Throughout August, the newspaper *La Nación* published a series of articles criticizing the decision to bring him to the city. An affronted General Vintter wrote a letter to the Ministry of War in which he argued that both warlike and 'tame' Indians were 'thieves and murderers', and that he was therefore committed to the 'extirpation of every savage from the theatre of his hunts'. Orkeke never saw this 'theatre' again. That same month, he caught bronchitis and was sent to the Military Hospital in Buenos Aires, where he refused to remain in bed, preferring instead to walk up and down and 'breathe the air ... free, as in my land'. In September, he died in the hospital. Eight days later, *La Nación* informed its readers that Orkeke's body was being marinated in water and lime in order to remove the flesh from its bones, and that the public would shortly be able to view his skeletal remains.

That same month, Orkeke's skeleton was exhibited in the Military Hospital, and it was subsequently transferred to the Museo de La Plata and presented to the public as the 'fossil representative of a race of giants'. Such practices cannot be understood purely in terms of their perceived value to science. In his study of nineteenth-century museum culture in Argentina and Brazil, *El museo vacío* (The empty museum, 2013), the cultural historian Álvaro Fernández Bravo has depicted the 'primitive accumulation' of objects and images in Latin American museums as an attempt to establish a 'national' heritage in newly independent countries that were still in the process of defining themselves in relation to each other and to the rest of the world.[11] If the Museo de La Plata's anthropological collections told a 'scientific' evolutionary story of the racial 'stages', they also told a patriotic story, of a modern progressive nation, dedicated to the acquisition of knowledge on the burning question of human origins.

Moreno, like Ameghino and Sarmiento, was a proponent of the 'American man' hypothesis that humanity might have originated in Patagonia and the Pampas. The museum's research priorities were always subject to budgetary constraints and the vagaries of national and regional politics, and the quality of its research sometimes left much to be desired. Moreno and his fellow-collectors often made assumptions about the ethnic or racial origins of their exhibits on the basis of where they were found, even though many of the peoples whose remains they collected were nomadic or semi-nomadic. These collections were poorly archived and inventoried. In 1889, Moreno complained to Zeballos that the labels identifying his donations had fallen off in transit and that he was unable to match the skull that supposedly belonged to Calfucurá to any of the skeletons he had been sent. Moreno was frequently criticized by his peers for his overbearing temperament and his methodological sloppiness. Víctor Grau-Bassas, the first inspector of the Museo de La Plata's library, called its director a 'rank ignoramus'. Ameghino fell out with his former colleague to the point that he publicly denounced Moreno in his bronze prize-winning study of Argentinian fossil mammalia at the 1889 World Exposition as a 'vulgar charlatan', whose museum 'offers not even the most minimal guarantee of seriousness, as it is in the hands of a megalomaniac who dreams and raves of grandeur'.[12]

* * *

Despite Ameghino's claim that Moreno was 'making our country ridiculous abroad', the museum attracted a stream of foreign scientists and visitors, some of whom were recruited by Moreno. In 1887, the American naturalist Henry Ward hailed its 'incalculable scientific value'. In 1892, the itinerant Dutch anthropologist Herman Frederik Carel Ten Kate Jr (1858–1931) met Moreno in Buenos Aires and was appointed head of the museum's anthropology section on two separate occasions. In 1897, his place was taken by the German anthropologist Lehmann-Nitsche, who remained in Argentina for thirty years at the Museo de La Plata and at the University of Buenos Aires. Lehmann-Nitsche was a specialist in the relatively recent discipline of 'anthropo-pathology'—the study of physical deformations, arthritis and other conditions in osteological remains—and presented skulls from the Río Negro to the German Anthropological Society as evidence that prehistoric peoples had carried out trepanning and other forms of surgery. Ten Kate sought to establish physical connections between the longer 'dolichocephalic' Indigenous skulls in Patagonia and the Indigenous peoples he had studied in the Pacific.

Photography was an important instrument of the museum's ethnographic and anthropological studies, and Indigenous subjects were photographed at the museum's studio and on field trips into the interior. Lehmann-Nitsche took a bespoke 'Bertillon chair' on his expeditions in order to standardize the distance and the measurements of his Indigenous subjects, who were photographed in frontal portraits and profiles. These photographs also became part of the museum's anthropological and anatomical galleries, along with busts of individual caciques made from wax and bronze. One wall of Moreno's anthropological gallery was lined with face masks illustrating different racial categories, above cabinets displaying foetuses in jars. Other display cases contained rows of skulls or complete skeletons, with labels explaining their type and, in the cases of the more well-known caciques, their names—a process that blurred the distinctions between objects of scientific value and trophies of conquest.

Today, the museum has been reconfigured in line with twenty-first-century expectations. Moreno's skulls and skeletons have either been removed or replaced with replicas. The museum's anthropological and ethnographic galleries contain a fierce-looking mannequin of Rosas's adversary Chocorí, wearing the thick guanaco 'armour' captured by Rosas's soldiers after Chocorí's escape from Choele Choel in 1833. According to the explanatory text, this display is intended for purely historical reasons to illustrate the 'particular exhibition forms of the nineteenth century and the beginning of the twentieth' in order to 'conserve the memory of the First Peoples of Argentina' as 'cultural mirrors'. Instead of displaying the physical relics of 'lost' and/or 'inferior' races, the museum now invites the visitor to 'reflect on the future, to value cultural diversity and recognize ourselves immersed in a pluricultural world' while also celebrating Argentina's First Peoples as 'heirs of a strong cosmovision profoundly connected to nature, the earth and the universe' through the process of 'recovering identities often hidden by the Western gaze'.

In its nineteenth-century iteration, the museum was very much designed for the Western gaze, by the man whose bust still stands in the main reception hall, surrounded by paintings of the 'lost races' of the Pampas. In these panelled walls and galleries, visitors could contemplate the relics of men and women who had once inhabited the conquered 'desert'—the links in a chain that connected the Conquest of the Desert to prehistorical extinctions. And some of these visitors may also have encountered the living men and women, who were also part of Moreno's evolutionary procession, in one of the strangest and saddest chapters in the Museo de La Plata's history.

16

'PRISONERS OF SCIENCE'

This story begins in 1885, when Moreno met the captured Mapuche/ Tehuelche caciques Inacayal and Foyel at the Tigre military barracks in Buenos Aires. Moreno knew both chiefs well from his Patagonian travels, and some of his journeys would not have been possible without their assistance. In an article in the newspaper *El Diario*, he described his sadness at the sight of his 'old friends from the desert' languishing in captivity at the military barracks:

> In the half-light of a room I see the men on one side, the women on the other. Inacayal is lying down; Foyel is squatting with his head bowed, he no longer has the untamed look that gave him the name of a fine warrior. All of them look dejected, and at first they don't recognize me, but seconds later the two of them get up at the same time, smiling. 'Moreno', they say, and reach out their right hands. At last the witness has arrived who will say 'We aren't bad Indians.' And they aren't, and they know that I know this.[1]

Moreno's sympathies were undoubtedly genuine, regardless of his description of his 'old friends' as relics of 'the birth of humanity, in its first fumbling days; those men still wrapped in hides; those half-naked, miserable, uncultured women, who evoked the hard geological epoch of the past, of our grandparents'. Inacayal and Foyel insisted they had done nothing to justify their detention, and Moreno agreed that among these prisoners 'there was not a single one who has maltreated a white man maliciously'. He told his readers that 'these poor Indians believe I have some weight' and promised to 'use all my influence to return them to the southern lands'. This did not happen. When his friends were moved to the prison-cum-concentration camp at Martín García Island, Moreno used his contacts to secure the release of the two caciques and their families on condition they take up residence in the basement of the Museo de La Plata. The terms of this arrangement were not clear. Máximo Farro, a professor of anthropology at the University of La Plata and the author of the gold standard history of the museum's first collections, has spoken of the 'great documentary vacuum' regarding the transfer of Inacayal, Foyel and their families to La Plata.[2] Moreno's male 'guests' were expected to perform manual tasks or act as nightwatchmen, while the women were given looms to weave ponchos, sashes and other textiles, some of which they sold in La Plata in exchange for alcohol, tobacco or sweets. The women may also have been expected to perform these activities for the benefit of the public.

Over the next few years, as many as two dozen Indigenous men and women from different parts of the country may have lived in the basement in equally nebulous circumstances, for purposes that were never entirely clear. In May 1886, Moreno told the minister of public works that he had begun to undertake a 'psychological study of the Argentinian tribes' based on 'living representatives of the most inferior races of Tierra del Fuego, a Yagán and an Alacaluf' who were staying at the museum.[3] These studies, Moreno claimed, would uncover 'many mysteries of prehistoric humanity, from the times of the childhood of primitive man' and the 'beginnings of human sociability'. In saving his former friends from hard labour, Moreno had also acquired a collection of living 'inferior races' in the same building where the remains of 'lost races' were exhibited that could be consulted, measured, photographed and studied without having to go out into the 'desert' in search of them.

* * *

Most of the available information on the museum's Indigenous residents comes from the Dutch anthropologist Ten Kate, who attempted to elicit ethnographic information from Inacayal and his companions, albeit without success. According to Ten Kate, the men did little work and spent most of their time smoking and drinking yerba maté or alcohol. They were often angry, morose and fractious, especially when drunk. At times, they lapsed into a state of 'apathy', which they expressed through 'extremely lugubrious' songs. Inacayal expressed 'the greatest sadness' and could often be found roaming the museum's corridors, exclaiming 'I am chief, you White thieves ... killed my brothers, stole my horses and the land where I was born, later [turned] a prisoner ... I am a miserable [creature].' An anonymous visitor described Foyel as 'a crowned head of the desert, who seems ill-matched with his new life in those cloisters, judging by the answers he gave to our questions'.[4]

Not all of Moreno's 'guests' were so despondent or uncooperative. The Argentinian criminologist and diplomat Luis María Drago (1859–1921) visited the museum as part of the research for his book *Los hombres de presa* (The men of prey, 1888). A follower of the 'criminal anthropology' school pioneered by Lombroso, Drago's discussion of Argentinian criminality contains various references to Darwin, the Fuegians and other Argentinian Indigenous Peoples. Drago believed that Indians were not incapable of civilization, and he based these arguments, in part, on his observations of the Indians at the Museo de La Plata 'recently brought from their *wigwams*'. Drago described one of the 'rude sons of the desert, with his bronze forehead, his thick black hair and his grotesque features, dressed in the clothing of civilization, adjusting the soldering of some fossil bone or preparing the embalming of a bird, in midst of windows, bottles and fragile collections'.[5] He also met a young Yagán man named Maish Kensis, who had arrived at the museum unsociable and 'even savage' but who had become 'timid, obedient and faithful' and a 'useful auxiliary for the Museum'.

According to Drago, Kensis showed a 'marked taste' for Western fashion. He was fond of 'strolling the streets of La Plata wearing a coat which Moreno had given him, perfumed and pomaded', and he played with Moreno's children. Such behaviour suggests that Moreno's Indigenous 'guests' enjoyed some freedom of movement within the limited options available to them. The museum's photographer Pozzi took a series of Bertillonage-style front and profile pictures of Inacayal, his wife Margarita, Foyel and some of their relatives and companions, and some of these pictures were exhibited in the museum. The Italian-born sculptor Victor de Pol (1865–1925) made various 'ethnic sculptures' at the museum, including a bust of a Fuegian woman named Eulltyalma or Tafá, who was 'rescued' from Martín García Island by Moreno and died in the museum on 9 October 1887. The bust was sent to Europe to be cast and exhibited, where one critic praised the 'crude and disturbing example of a human being so distinct from our civilized humanity', in whose 'simian features' the 'muscles found in the superior races that produce movement in the nose' were 'completely atrophied'.[6]

Eulltyalma was one of a chain of Indigenous deaths at the museum. Foyel's wife Margarita died on 23 September that year, followed by an unnamed seven-year-old girl two days later. To the astonishment of the local press, all three were immediately dissected and soaked in quicklime to remove their flesh, and their bones were labelled and displayed in the museum's anthropological galleries. One local newspaper put in a formal complaint, calling for clarification of the status of 'the poor Indians, destined to die for the Museum', and asking why the police and coroner had not been involved in the aftermath of their deaths. Moreno responded in an open letter that the Indians were at the museum to facilitate linguistic and ethnographic observations regarding 'races that were very difficult to study in their savage habitats', and that they had died of tuberculosis and other pulmonary diseases acquired during their transportation from Patagonia to Buenos Aires. He insisted that all three had received medical treatment and justified their dissections on the basis of their exceptional interest to science 'as the last representatives of disappearing races'. He also claimed he had made a formal application to the Department of Hygiene to allow him to dissect any future remains 'necessary for the study of Patagonian indigenous peoples and of other races which may shed light on the knowledge of our origins'.[7]

* * *

One of these dissections was performed on the cacique Inacayal, who died in the museum on 26 September 1888. According to Moreno's secretary, Onelli, Inacayal sensed his death and prepared for it beforehand. In language steeped in the 'last wild Indian' romantic pathos that was often attached to the nineteenth-century's 'vanishing races', Onelli recounted the last day on earth of the 'powerful Araucanian cacique, taken prisoner in the war of the desert, [who] lived freely in the Museo de La Plata; there he hardly moved from his old man's chair'. According to Onelli, two Indians carried Inacayal to the top of the museum stairway, where he removed the clothing imposed on him by the 'invader of the homeland' and stripped to his waist. After bowing to the sun and to the south, Inacayal muttered an invocation, whereupon the 'old Lord of the earth disappeared like the rapid evocation of a world' and died later that night 'perhaps content that the victor had allowed him to salute the sun of his native land'.[8]

Onelli does not say how Inacayal died, though Ten Kate later examined his skull and claimed that his nose was broken and he was missing various teeth—an outcome suggesting an accident or suicide. Whatever the cause of his death, Inacayal joined his wife in the museum vitrine, where their skeletons were exhibited as examples of the 'ancient lords of the pampa'. Moreno was not displeased at the

> deaths of four Indigenous adults in the establishment [which] ... have given the Museum four brains of great scientific importance as the only members of this race that are conserved in collections, and four skeletons ... These pieces can be considered unique in every sense, and their importance increases when we take into account that they belong to individuals from a race that is rapidly being extinguished, and that within a few years can be called lost.[9]

Not all of Moreno's 'guests' ended up in the museum. Foyel and his family were able to return to Chubut, and others were also allowed to leave. It was not until 19 April 1994, on the 'Day of the American Indian', that Inacayal's skull and skeletal remains were flown to the city of Esquel, where they were greeted by representatives of Patagonia's First Peoples, including his grand-daughter Mercedes Nahuelpán. The box containing the chief's remains were draped in his poncho and the Argentinian flag and driven 59 miles to the town of Tecka, more than 1,000 miles from Buenos Aires, where they were buried in a mausoleum, surrounded by offerings of flowers and stones.

In 2004, a second restitution was required when Inacayal's brain, scalp and death mask were discovered in the museum laboratories, and these remains were also buried in the mausoleum. On a warm autumn day, I drove from Esquel to Tecka to visit Inacayal's final resting place, overlooking the mighty Route 40 that runs parallel to the Andes from the north of Patagonia to the south. I climbed up the hill, past the information boards that told the story of Inacayal's life and death and sketched out the history of the Conquest of the Desert until I reached the stone-embossed igloo-like tomb with the conical tin roof. Immediately below the hill, beyond the highway and the YPF petrol station, I could see the neat rows of houses and yellowing cypress trees and the brown wall of the Andean foothills in the background.

To the north, west and south, the immense plains stretched out into the far distance, broken by giant outcrops of rock and flat-topped mountains, and apart from Tecka there was no sign of human habitation. This was once the country of Inacayal's *toldos*. In 1879, Moreno came here, accompanied by Inacayal's son Ultrac. Upon their arrival, Inacayal and his warriors rode out towards them, yelling and stabbing the air with their knives and lances in order, as Moreno put it, to chase away 'the evil spirit that might engender distrust amongst the visitors and their hosts'. During the long debate, Moreno tried to reassure Inacayal that his intentions were peaceful. He called his hosts *paisanos* (countrymen) and hailed the Argentinian flags on their tents as proof that 'we were all Argentinians and we all had the same government in Buenos Aires'.

Moreno may have believed these words when he said them, but at that time the government in Buenos Aires did not see the 'barbarians' of Patagonia as fellow-citizens, let alone as rightful owners of the lands they occupied. This was why Inacayal was later hunted down and brought to Buenos Aires as a prisoner. It was not until his remains were finally returned to Tecka, more than 100 years later, that he and his people were recognized, as one of the information boards put it, as members of the 'First Peoples of Patagonia [who] loyally served the Argentine State during the 19th century and contributed to the consolidation of our frontiers.'

* * *

Inacayal is one of a number of Indigenous restitutions from the Museo de La Plata that have taken place in recent years, most of which date back to the late nineteenth century. In 2001, the remains of Rosas's adopted godson Mariano Rosas/Panguitruz Guor were returned to his community in Leubucó. In 2016, the remains of four Mapuche-Tehuelche chiefs were returned to the same community in Trenque Lauquen. In 2018, the partial remains of Moreno's Tehuelche friend Sam Slick were returned to the province of Santa Cruz, followed by a further restitution three years later. In 2018, the skull of Cipriano Catriel—Moreno's prized possession—was returned to Azul from the Moreno museum and buried with his poncho in a monument overlooking the territories once occupied by his *toldos*.

Other restitutions are ongoing or waiting to be resolved. The first restitution of Inacayal was a pivotal moment in the restitution process that paved the way for the legal changes required to formalize the return of Indigenous remains to the communities that claimed them. But the later return of Inacayal's 'soft parts' was the result of an investigation carried out by the Colectivo GUIAS (Grupo Universitario en Investigación en Antropología Social (University Investigation Group in Social Anthropology)), a group of anthropologists and former students from the Anthropology Faculty at the University of La Plata that has played a key role in the restitution process. The collective was formed in 2006 by staff and students who began to apply *antropología crítica* (critical anthropology) to the identification and conservation of Indigenous remains at the university's Faculty of Natural Sciences and the Museo de La Plata. In the course of these investigations, the group discovered that Inacayal's remains had not been fully restituted, and that the skeleton of Kensis was still exhibited in the medical faculty as a generic example of a 'human skeleton'.

The GUIAS anthropologists began collating information on other Indigenous 'prisoners of science'—an expression the group uses not only to describe 'living exhibits' at the Museo de La Plata but the remains of the dead—which they made available to the communities these remains came from. I spoke to Fernando Miguel Pepe Tessaro, one of the group's founder-members, and its photographer Marcos Bufando Fernández, at the Escuela Superior de Mecánica de la Armada (ESMA) or Higher Navy Mechanical School in Buenos Aires—the most infamous of the torture and murder centres created by the 1976–82 dictatorship—where the

National Institute for Indigenous Affairs (Instituto Nacional de Asuntos Indígenas; INAI) also has its offices. This location is not accidental. Today, the ESMA site has become a museum and site of remembrance for Argentina's 'disappeared'. The GUIAS collective sees its own practice of 'critical anthropology' as a tribute to the fifty-eight students and lecturers from the University of La Plata who were murdered or disappeared by the dictatorship, nineteen of whom were members of the Anthropology Faculty.

Pepe is a tireless advocate of Indigenous rights who has been associated with many of the restitutions of the last two decades. As is often the case on the Argentinian left, he and his fellow-students first began to look at the issue of Indigenous restitutions at the Museo de La Plata following encouragement from the ubiquitous Osvaldo Bayer. 'We quickly learned that there had been a group of Indigenous People inside the museum as prisoners,' he said. 'We learned about the restitutions of Inacayal and Mariano Rosas. We saw there was a kind of invisibilization of this group, because they were First Peoples, because of class, and also because of gender— there were women whose remains should have been restituted in 1994 and they weren't.' When Pepe and his colleagues began their investigations, Indigenous skeletons were no longer exhibited, but their remains were still stored in the museum and were often poorly catalogued and identified. The collective began to compile its own documentary and photographic record of the museum's Indigenous collections, and this data was then made public and opened the way for further restitution claims.

* * *

Both the restitution process and the 'critical anthropology' practised by Pepe Tessaro and his colleagues reflect a very different understanding of race and racial science from the norms that once shaped the Museo de La Plata's collections. How has the contemporary Museo de La Plata responded to this transformation and the ethical and logistical challenges it entails? And how does it regard the science and the ethics of its founders? I discussed these questions with the museum's director Dr Analía Lanteri and Professor Gustavo Barrientos, head of the Anthropology Department at the

University of La Plata. Dr Lanteri traced the changes that have taken place from a time when 'distinct peoples were being studied for their distinct anatomical, physiological and cultural characteristics' to the more recent notion of 'ethical, informed consent' based on 'the recognition that when you are doing research you are working with human beings' to whom researchers owed respect.

Such respect was largely absent from Moreno's anthropological collections, and Professor Barrientos dates the museum's change of attitude to the 1930s, when the physical remains of Indigenous Peoples were withdrawn from public exhibition and distributed between different university departments. It was not until 1991 that National Law 23.940 authorized the restitution of Inacayal. In 1994, Article 75, sub-section 17 of the new post-dictatorship Argentinian constitution officially recognized the 'ethnic and cultural pre-existence of the Argentine Indigenous Peoples' for the first time. This new article effectively replaced the 1853 constitution's commitment to 'preserve peaceful relations with the Indians and promote their conversion to Catholicism', and it was enacted in close consultation with the representatives of Argentina's First Peoples.

The recognition of Indigenous 'legal personhood' facilitated the process of restitution. In 2001, the Argentinian Congress approved Law 25.517 on the Restitution of Aboriginal Mortal Remains, which established that Indigenous remains 'that form part of museums and/or public or private collections should be placed at the disposition of the Indigenous peoples and/or communities of origin that claim them'. It was under this dispensation that Inacayal's 'soft parts' were returned to Tecka in 2005. Communities seeking restitution have first to be recognized by Argentina's INAI, before submitting a written request for restitution, which is sent first to the University of La Plata's Faculty of Anthropology and then passed on to the museum. The solicited remains are re-examined to confirm their identity, and the university then organizes a meeting with representatives of the community seeking restitution. Once the restitution has been signed off, representatives of the community come to the museum, bringing musical instruments, and another ceremony is performed when the remains are actually buried.

This is what happens in the best of cases. But there is considerable room for error in Moreno's chaotic collection. Sometimes names are listed without any corresponding remains. At other times, the different parts don't match, or modern scientific methods have found them to be older or younger than they should be. There are also cases in which different Indigenous communities make claims and counter-claims for the same remains. According to Lanteri, some communities regard the museum, and scientists in general, as 'the enemy' and refuse to accept their priorities. 'There are people who say you shouldn't ever study human remains, that it's a lack of respect,' she said:

> The museum doesn't take in any more human remains, but the Faculty of Medicine does. In the La Plata cemetery, there are remains in common graves that go to the Faculty of Medicine, for the study of osteoporosis. Some remains are not historical or archaeological. They might be millions of years old. These remains are very important for establishing human antiquity.

After our conversation, Professor Barrientos showed me the basement where Inacayal and his companions once lived, and where some of them died. Today, this part of the museum is mostly taken up by laboratories, offices and workshops, but Barrientos took me to the storeroom, where Indigenous remains awaiting restitution are kept in white cardboard boxes. Only one of the boxes was open, and a pair of remarkably well-preserved mummified infants lay facing each other as if they were embracing. Their bodies were some 6,000 years old, and they had preserved so well in the soil that their reddish hair and facial features were still visible. The other boxes were only identified by labels such as the 'wife of the cacique "Sapo"', 'Maish Kensis'—the young Fuegian dandy who lived and died at the museum—and a woman listed only as 'María'.

My attention was immediately caught by the box labelled 'Calfucurá'. The origins of these remains had been questioned by Moreno himself, and their restitution had been equally problematic because Calfucurá had many wives and children in many different places and a number of communities have claimed his remains. Nor was it certain that the box actually contained the remains of the great war-chief. Examinations suggest that these bones and skull belonged to a man in his thirties rather than the eighty-year-old whose remains had supposedly been dug up by soldiers in Carhué and given to Zeballos for his collection. It remains to be seen how these and other restitutions will be resolved as the museum seeks to reconcile Moreno's flawed legacy with the claims of the communities for the restitution of these 'prisoners of science'.

* * *

Today, anthropologists no longer see 'primitive' cultures as relics of disappeared or disappearing races, and the Argentinian state is attempting, however slowly, to recognize its Indigenous Peoples as equal citizens in a multi-ethnic society. Nevertheless, Moreno's 'prisoners of science'—and their close associations with the Conquest of the Desert and other nineteenth-century military campaigns—have left a legacy of suspicion and distrust that is not easily overcome. The Arroyo Seco 2 archaeological site, near the city of Tres Arroyos in southern Buenos Aires Province, is a recent example of these tensions. In the 1970s, archaeologists uncovered what appeared to be a human burial ground-cum-hunting site containing seventeen human skeletons. More recent excavations by a team of archaeologists from the Universidad Nacional del Centro de la Provincia de Buenos Aires have found that these remains are between 6,000 and 14,000 years old. So far, forty-four human skeletons have been uncovered at the site, in addition to the remains of giant sloths, giant armadillos, camels, deer, guanaco and larger mammals. These are some of the oldest human remains ever found in Argentina, and they transformed Arroyo Seco 2 into one of the most important archaeological sites in South America.[10]

Some Indigenous communities have described these excavations as a violation of their spiritual traditions and cosmology. In 2021, a local association called the Associación Encuentro Indígena (Indigenous Meeting Association) solicited INAI to halt them and return the original remains found at the site, basing their claim on Law 25.517. The archaeologists overseeing these excavations argue that these remains are too old to have any connection to any living Indigenous community. When I visited Tres Arroyos in 2023, the dig had been halted, and I discussed the controversy with Evangelina

Balcedo and Alicia König, whose association has been campaigning for more than three years to transform the Arroyo Seco 2 site into a space of Indigenous 'remembrance'. How did they reconcile these demands with the absence of any clear links between the human remains found at Arroyo Seco and any present-day Mapuche community?

Evangelina admitted that it was impossible to say whom these remains belonged to, and she insisted on the need to establish a 'balance' that transcended the question of their antiquity or their origins. At present, this balance has been achieved through the site's reconstitution as a commemorative space that recognizes the 'pre-existence of Indigenous Peoples'. Did there have to be a conflict between science and restitution? I asked. 'Today, science can determine from 20 grams of collagen what these people ate, how they lived, lots of specific data,' Evangelina replied:

> For them, it's not enough. They want to go on investigating and investigating indefinitely, and at any moment they might want to keep these remains in some other place. If a community agrees that in the interests of science they should go on investigating, fine. But not if they do it without any consultation with people. There needs to be another kind of dialogue. And if it doesn't happen, then respect doesn't exist.

This question of respect is crucial to the restitution process and to the indignation felt among Argentina's First Peoples towards Moreno's collections. Evangelina and Alicia took me to the Arroyo Seco 2 site with some other members of the association, including its Mapuche president Patricia Cayulao. The diggings can only be viewed from a wooden observation tower, and we stood at the edge of the fenced-off field, beneath the sign declaring the site a 'Site of Remembrance of First Peoples' while Patricia knelt on the ground and let out a haunting prayer/incantation that Evangelina explained was a request to her ancestors for permission to enter the site. Permission was granted, and as I looked down from the tower at the excavation site, I wondered what else it contained, and how anyone can know if the 14,000-year-old human remains found in these diggings were their ancestors.

On one hand, the question is irrelevant, because Arroyo Seco is exactly what Evangelina and Alicia said it was: a symbolic conflict about the visibility, respect and recognition towards Argentina's Indigenous Peoples that was rooted in the 'pillaging science' of Moreno and Zeballos and the historical revisionism surrounding the Conquest of the Desert and its consequences. Twenty-first-century scientists are no less curious than Moreno about the origins of 'American man', but now they are obliged to conduct their research according to an entirely different set of ethical norms. This is why even palaeolithic remains have become—literally and figuratively— bones of contention in the ongoing search for recognition and respect that was conspicuously absent when Argentina's vanquished 'savages' became the raw material for Moreno's dubious science.

17

FIN DEL MUNDO

On 14 July 1887, an unusual banquet was held at the Star and Garter Hotel in Richmond, London, in honour of the former president of Argentina, Julio Argentino Roca. City bankers, investors and traders and distinguished lords and ladies all gathered to pay homage to the general who had completed his first presidential term the previous year. No Latin American leader in British history had ever received such a reception. In a long hagiographic piece that was published as a book, *La Tribuna Nacional*—the Roca's administration's propaganda mouthpiece—hailed Roca's presence among the distinguished representatives of 'the most advanced civilization' in the world as proof of Argentina's new status as a great nation. *La Tribuna* meticulously described the London neighbourhood, the menu and seating arrangements, the wealth and social rank of the men, the elegance of the women and the musical pieces played by the band as a testament to the greatness of the English—the better to magnify the honour bestowed on Roca and on Argentina.[1]

Almost everyone present was an investor or potential investor in a country that was now seen as a profitable repository for British capital. In a keynote speech, John Baring, 2nd Baron Revelstoke, a senior partner in Barings Bank and a future director of the Bank of England, cited British investments in Argentinian railways, banks and trams as evidence of the deep ties of friendship between the two countries. Frank Parish, the son of the former consul Sir Woodbine Parish and a shareholder in the British-owned Buenos Aires Great Southern Railway Company, praised Roca's 'new conquest of a vast territory that formed part of the old discoveries and had remained since then in the power of the savages' and had created an 'immense zone of fertile lands that are today for the most part occupied by railways and in the hands of industrious settlers'.

This banquet was a confirmation of Argentina's new place in the world. Between 1870 and 1900, the value of Argentinian wool exports increased by 500 per cent. In 1830, Argentina had 2.5 million sheep and exported 6 million pounds of wool. In 1883, it had 69 million sheep and exported 261 million pounds of wool. Before 1870, Argentina imported wheat to feed its population. By the end of the nineteenth century, it was the fifth largest exporter of wheat in the world, exporting an average of 1 million tons annually in the 1890s. Between 1884 and 1908, the Argentinian railway network increased from 5,745 miles of track to 15,476— most of which was owned by British companies. Britain provided Argentinian ranches with new breeds of sheep and cattle—designed to please the British palate—in addition to barbed-wire fencing, freezer lockers and refrigerator ships. Immigration, as Sarmiento and his peers had hoped, was a decisive factor in Argentina's economic transformation. In 1869, the population of Argentina was 1.8 million. By 1895, it was 4 million, and by 1914 it had doubled again to reach 8 million.

This dizzying ascent reached its apotheosis at the 1889 World Exposition in Paris, where the opulent Argentine pavilion next to the Eiffel Tower attracted more than 32 million visitors. The *Guide bleu du Figaro* described the pavilion as a symbol of the Argentine Republic's 'immense pastures, and its colossal export of cattle'. Exhibits of refrigerators, saddlery, canned food, frozen meat and woollen products demonstrated Argentina's prosperity, technological prowess and productivity. Moreno's former collaborator at the Museo de La Plata, Ameghino, won a bronze medal for his voluminous *Mammalian Fossils in the Argentine Republic*. The Argentine pavilion at the previous World Fair in 1878 had featured Moreno's photographs of Indigenous skulls. This time, there was no reference to Argentina's Indigenous Peoples in the pavilion. When a French visitor asked one of the pavilion staff to see photographs of savages, she was told that such people no longer existed, except those who 'had moved to the cities', who were now 'educated, valued citizens'.[2]

So concerned were Argentina's rulers to eliminate any trace of its Indigenous past that the Argentinian vice-president Carlos Pellegrini complained to his brother that most of the uniformed ceremonial guards outside the pavilion's main entrance were mixed-race Argentinian 'chinos' with Indigenous features rather than the more obviously white representatives whom Pellegrini regarded as 'the true figure of our soldier'.[3] In effect, the pavilion symbolized the Argentina that the Generation of '37 and their successors had aspired towards: a modern, progressive country based on mass European immigration, foreign capital and agricultural exports—a country of culture, science and technology in which the 'desert' had been made productive, and its 'savage' inhabitants had either vanished or become 'educated, valued citizens', even in the territories where Darwin had first seen them.

* * *

In a travel piece on Argentina for *Harper's Weekly* in November 1887, the American writer William Eleroy Curtis noted that there 'used to be a place called Patagonia' into which 'ranchmen have been pushing southward with great rapidity, and now the vast territory is practically occupied'. As a result, '[t]here are more cattle or horses there than in Kansas, and the dreary, uninhabited wastes of Patagonia have gone into oblivion with the "Great American Desert"'. Though 'the remnant of a vast tribe of aborigines still occupies the interior', Curtis informed his readers that 'the Indian problem of the Argentine Republic was solved in a summary way', and the frontier cleared of its 'bands of roving savages' by 'General Roca, the Sheridan of the River Plate, [who] was sent with a brigade of cavalry to prevent this sort of thing.'[4]

These claims were overstated. Patagonia did exist, even if it was no longer what it used to be. In 1877–8, the disputed claims between Argentina and Chile over Patagonia brought the two countries to the brink of war. In 1881, this dispute was resolved through negotiations in which Argentina conceded the Strait of Magellan and half of Tierra del Fuego to Chile, while Chile relinquished its claim to mainland Patagonia east of the Andes. The final agreement still left the exact state border along the Andes undefined, and these limits were eventually decided by international arbitration in 1902. But the 1881 treaty—coupled with the fact of Argentinian military occupation—ended Patagonia's disputed status and paved the way for the region's absorption into Argentina's agricultural economy.

This process was initially restricted to mainland Patagonia as a result of Roca's military campaigns. Further south, colonization took a very different form. Today, some 400,000 tourists cross the Sierra Alvear to visit the city of Ushuaia ('deep bay' in the Yagán language), overlooking the Beagle Channel at the edge of the Isla Grande in Tierra del Fuego. Some come to walk in the mountains or forests, while others take boat trips in search of sea lions, whales and penguins or have their photographs taken by the sign overlooking the bay that proudly proclaims 'Ushuaia: Fin del mundo'—Ushuaia: End of the world. Ushuaia is not the most attractive city. The philosopher Jean Baudrillard harshly condemned its 'chaotic, incoherent cowboy-film modernity' and its 'concrete, dust, duty-free, transistors, petrol, computers and the hubbub of useless traffic' as manifestations of 'civilization's wastes'.[5]

Apart from its splendid natural setting and its unique location at the 'end of the world', Ushuaia's main attraction is its heavily Darwinized history. Restaurants and hotels bear names like 'Jeremy Button [sic]' and the Hostal Yaghan; a 'Thematic Gallery' diorama contains life-sized figures of FitzRoy, Darwin and the Yagán people. At the Beagle Centre, actors on a life-sized replica of the *Beagle* act out a musical multi-media show called 'The Adventure of the Beagle' that enables the visitor to 'enter the Patagonia Story that changed the World'. After the show, punters can 'get a live experience of the southernmost aboriginal culture' by sitting around stone tables with electric fires, using harpoons as forks and mussel shells as plates.

It was here, in 1869, that the catechist Waite Stirling established the renamed South American Mission Society's first permanent mission in Tierra del Fuego. In 1871, the twenty-eight-year-old Thomas Bridges, a catechist at the Keppel Island Mission, came to live with Stirling, along with his wife and daughter, in the prefabricated hut supplied by the society. When Stirling left the mission to take up his duties as bishop of the Falklands in Port Stanley the following year, Bridges continued the work that he had begun. Bridges had already begun to learn Yagán on Keppel Island and became the first white man who could actually communicate fluently with the Yagán people in their own language. Under his determined management, Ushuaia became a model 'Christian village', with roads, houses, a school and an orphanage, where Yagáns wore clothes and lived in wooden huts, where they cultivated gardens and bred goats and cattle. Following the Chile–Argentina boundary negotiations in 1881, Ushuaia became part of the Argentinian half of the Isla Grande. In September 1884, four Argentinian navy ships arrived to establish the new sub-prefecture of Ushuaia, and the Argentinian flag was flown at the mission station for the first time.

This was a historic moment for Argentina, but the expedition also caused an outbreak of measles and tuberculosis that killed more than half the Yagán population at the Ushuaia mission within a few weeks and another 50 per cent of the survivors over the next two years. It was partly in response to this catastrophe that Bridges travelled to Buenos Aires in 1886 to apply for a grant of land some 40 miles east of Ushuaia, on which he proposed to 'create a farm, and employ native labor upon it'. With the support of Moreno and President Roca, Bridges was given 200,000 hectares overlooking the Beagle Channel to establish a farm that he claimed would 'ensure the well-being of some of the natives'.[6] The South American Mission Society took a dim view of what it regarded as a purely commercial enterprise and disavowed the project. In April 1887, the Bridges family moved into the new settlement, which they initially called Downeast before later renaming it Harberton after the Devon village where Bridges' wife was born. The family were accompanied by sixty-two Yagáns from Ushuaia, who built roads, cut down trees and sawed the wood for the estate's white frame main house and for its outbuildings and fences.

The settlement quickly became a profit-making sheep ranch, which exported wool to England and sold meat to local miners and prospectors, and it also became an essential port of call for scientists and sailors passing through the region. One of them was President Roca, who visited Harberton in 1899 during his second term, while returning from a diplomatic meeting with his Chilean counterpart at Punta Arenas. In his classic memoir-cum-adventure narrative *Uttermost Part of the Earth* (1948), Bridges' son Lucas describes how Roca and his party were served 'strawberries and cream in real Devonshire style' made by his mother—the ultimate confirmation of his late father's efforts at transforming 'irresponsible savages into a law-abiding community' with a 'keen sense of law and order and of property rights'. Bridges had grown up among these 'irresponsible savages' as a child, and as an adult he regarded them, much as his father did, with a mixture of respect and paternalistic affection. A Yagán speaker himself, he criticized Darwin for his 'shocking mistake' in describing the Yagáns as cannibals—an error he attributed to the eagerness of FitzRoy's Fuegian captives to provide their interrogators with a 'delectable fiction'.[7]

Bridges also rejected Darwin's dismissal of the Yagáns as inarticulate brutes and cited his father's 'dictionary of the speech of Tierra del Fuego' as evidence of the expressive range of the Yagán/ Yamana language. Completed in 1879 and published posthumously by his children, the dictionary's 32,000 words and inflections included more than fifty words to describe family relations, five names for 'snow' and even more for 'beach' depending on the direction of the speaker and various other factors. Words like *sauiya* ('in a state of turmoil as a surging sea, surges, billows'); *miing-enasinana* ('To make believe one is fearless when otherwise'); *atal-agu* ('To jump, leap, spring, as a delighted baby in his mother's arms. To jump, leap, pulsate as the heart'); and *Mamihlapinatapai* ('to look at each other, hoping that either will offer to do something, which both parties much desire [dare] but are unwilling to do'), all suggest a range of emotional responses and social relationships that are entirely absent from Darwin's reductionist descriptions of the 'shivering hideous savage'.[8]

In 1886, Thomas Bridges wrote a letter to the Buenos Aires English-language newspaper *The Standard* that also contained an implicit rejection of Darwin in its claim that

> the language of one of the poorest tribes of men, without any literature, without poetry, song, history or science, may yet through the nature of its structure and its necessities have a list of words and a style of structure far surpassing that of other tribes far above them in the arts and comforts of life.[9]

Bridges had once answered Darwin's questionnaire on the expression of emotions among the Fuegians of Keppel Island, but his defence of Yagán culture was very much at odds with Darwin's earlier assessment of their cultural and social practices. In the notes and diary entries collected by his children after his death, Bridges described cannibalism as 'an utter impossibility from the very nature of native society, in which human life is a very sacred thing'. Infanticide was almost non-existent, save for a few exceptions carried out by women, usually in revenge for rejection or abandonment, since 'mothers of young children tend them most carefully, and give themselves up to their maternal duties much more entirely than is the case with most civilized people, scarcely ever putting them out of their arms, and nourishing then amply to a very late age'.

Bridges also described burial rites in which the bereaved cropped their hair and painted their faces with charcoal and oil, and even friends of the bereaved painted their faces red in sympathy. Where Lubbock, Tylor and so many other writers depicted savagery as a state of misery, fear and wretchedness, Bridges commented on the 'ease and sociability' of Yagán culture, which was

> eminently favourable to talk ... The natives among themselves gathered round the wigwam fire, their wants generously supplied (for as a rule they had abundance of food), were very animated and spent a large part of their time in lively conversation and exuberant merriment, laughing as much to please others as to express their own pleasure.[10]

* * *

This appreciation of Yagán culture did not mean that Bridges wanted to preserve it. If he rejected some of the misconceptions that had once defined the Yagáns as savages, the mission was as committed to their cultural and spiritual transformation as FitzRoy had been, regardless of its consequences. By the end of the nineteenth century, the Yagán population that may have numbered 3,000 when FitzRoy and Darwin arrived in Tierra del Fuego had dwindled to less than 400, largely as a result of diseases such as smallpox, measles, TB and typhus brought by outsiders. In May 1882, Bridges informed the society that the rate of mortality from these diseases was higher among the Yagáns who stayed away from the mission than among those who had adopted the 'clothes and civilized forms' and diet that he regarded as more beneficial to the natives. Yet that same year, Paul Daniel Hyades, the doctor on the French expedition to Cape Horn, treated some of the sick Yagáns at the mission and noted that 'pulmonary tuberculosis is rare amongst the Fuegians who live in the open air; but it is very frequent amongst those who inhabit the English mission at Ushuaia, who have acquired sedentary habits and live cooped up'.[11]

In a visit to Tierra del Fuego in 1894, the American journalist John Randolph Spears expressed his dismay at the 'Cape Horn Aborigines' who 'have been almost exterminated by changes in their habits, wrought by Christian missionaries'.[12] Spears attributed the decline of this 'hard and vigorous race' to the clothes they were obliged to wear, which left them more vulnerable to colds and other diseases than when they had lived naked in the open. He accused the missionaries of showing more concern for the Yagáns' spiritual salvation than for their physical well-being and quoted from a diary entry on an undated epidemic, which recorded: 'We have now lost forty-three persons in three weeks at Ushuaia. How far it has spread I cannot say. It has been a pleasure to go among them, for in almost every house we have heard the voice of prayer and praise in midst of all their sufferings.'[13] In 1883, Bridges met the aged Yokcushlu/ Fuegia Basket 'in a dying state' on Cook Island. In his diary, Bridges describes how he 'endeavoured to lead her to hope in God. I told her that many felt a great interest in her and would be sorry to hear of her death.'[14]

Bridges does not say whether he succeeded in bringing FitzRoy's former captive to God, but this outcome was the ultimate validation of the missionary enterprise in the 'uttermost end of the earth' as it was in other parts of the world. In 1888, the South American Mission Society established a new station at Bayly Island, near Cape Horn, headed by Leonard and Nellie Burleigh, two missionaries from the Keppel Island station. The willingness of a young English couple to move with their newborn child to what was effectively the southernmost Christian mission in the world and live among 'savages of the wretchedest type' was a testament to the profound religiosity, commitment and courage of the society's recruits. By February 1890, the Reverend E.C. Aspinall reported that the Burleighs had transformed 'the purest savages' into 'a quiet, docile, clean people'.[15] The mission ended tragically when Michael Burleigh drowned in a boating accident, and Nelly Burleigh returned to England. In a public talk, she recalled her first sight of 'one of the oldest men at Cape Horn', telling her audience: 'I am sure if you could have seen him you would have thought the same as Darwin did, and asked the question, as I did on three occasions, "Are these people really human?"'[16] That the same man died shortly afterwards 'rejoicing in Christ', in 'the first Christian burial that we had at Wollaston Island', provided an affirmative answer.

Work was an essential tenet of the 'arts and good manners of civilised life' that the missionaries brought with them to Tierra del Fuego and was generally rewarded not with money but with food and clothing. This was not slavery—the Yagáns lived in Ushuaia and Harberton by choice. But they worked twelve-hour days, loading and unloading boats and cutting firewood in return for a meagre diet of oatmeal, tea and biscuits, vegetables and scraps of food and some hunted seals and otters for the missionaries, who paid them with tea and biscuits and then sold the skins at a profit. Spears believed that Bridges exploited the Yagáns and predicted—correctly—that 'his cheap labour—the cheapest, for the purpose, found anywhere—and his ready access to market' would transform Bridges into 'one of the wealthiest land-holders in the south part of the continent'.[17] In 1922–3, the American graphic artist Rockwell Kent described Harberton as 'an immediate stepping stone from godliness to wealth', and he also claimed that the 'pestilence of Christian mercy' had turned the shores of Tierra del Fuego into a 'solitude'.[18]

This 'solitude' caught the attention of other visitors to the far south. When the American priest, anthropologist and sociologist John Montgomery Cooper visited Tierra del Fuego in 1917, he estimated that there were only 100 Yagáns left in the islands, most of whom lived in the Beagle Channel. Within a few decades, a society that had managed to survive in one of the most hostile and intemperate environments on earth for more than 7,000 years had almost vanished, at least in its original form. On the surface at least, Darwin's Fuegian 'savages' had joined the nineteenth-century's 'vanishing peoples' through a combination of European-borne diseases and a colonial-missionary enterprise that saw the disappearance of outward manifestations of savagery as a mark of civilizational and religious achievement. And all this could be traced back to the arrival of a British surveying ship in 1829 in the channel the Yagáns once called Onashaga and the decision of its impetuous captain to take four captives back to England.

* * *

The question of whether Darwin ever revised his initial impressions of the 'Fuegians' has often been a subject of historical debate. Some scholars cite the lifelong correspondence between Darwin and Sulivan, his former *Beagle* shipmate and future British admiral, as evidence that Darwin changed his opinion about the capabilities of Tierra del Fuego's 'savage' inhabitants. Sulivan was one of the original committee members of the Patagonian Missionary Society, and he persuaded Darwin to subscribe to the society and provided him with regular updates on the society's work throughout Darwin's life. In December 1866, he described a meeting with the four 'Fuegian lads' who had recently travelled to England with Stirling. One of them was Button's son, whom Sulivan described as 'a very nice lad & good disposition—but not so intelligent as one of them from Packsaddle Bay. One can hardly believe that this lad was the same race as those we saw alongside in that Bay where we taught them to rub their noses and say "Old Stokes".'[19] Darwin was pleased by this positive assessment. 'Your letter has interested me exceedingly all about S. America & the Fuegians,' he replied. 'I never thought the latter cd have been civilized, but it appears that I shall be proved wrong. I wish poor Fitz-Roy was alive to hear the result of his first attempt for the civilization of the Fuegians.'

In June 1870, Sulivan reported that Stirling had spent six months living alone with some 'friendly Natives' in Ushuaia, who were 'quite alive to growing potatoes &c and taking care of their goats. and one who with his wife lives at the Murray Narrows has a good garden & goats &c. they keep their house as neat and clean and themselves and children as nicely clothed & clean as Europeans'.[20] Once again, Darwin responded positively. 'I had never heard a word about the success of the T. del Fuego mission,' he wrote. 'It is most wonderful, & shames me, as I always prophecied utter failure. It is a grand success.'[21] In March 1881, Sulivan wrote a long letter to Darwin describing the progress of the Ushuaia mission, in which he described Bridges' dictionary and the missionary's conclusions regarding the 'perfect character' of the Fuegian language and gave some examples.[22] Darwin wrote to Sulivan:

> It was very kind in you to answer & interest not only me but all my family by your very curious account of the Fuegians. It is truly wonderful what you say about their honesty & their language. I certainly shd have predicted that not all the Missionaries in the world could have done what has been done.[23]

In December that same year, in the last months of his life, Darwin sent Sulivan a late payment of 2 pounds towards his subscription in a letter that concluded: 'Judging from the Missionary Journal, the Mission in Tierra del Fuego seems going on quite wonderfully well.'[24] Looking back on these exchanges in a letter to the *Daily News* on 4 April 1885, Sulivan recalled:

> Mr Darwin had often expressed to me his conviction that it was utterly useless to send Missionaries to such a set of savages as the Fuegians, probably the very lowest of the human race. I had always replied that I did not believe any human beings existed too low to comprehend the simple message of the Gospel of Christ. After many years ... he wrote to me that recent accounts of the Mission proved to him that he had been wrong and I right in our estimates of the native character and the possibility of doing them good through Missionaries.[25]

Darwin's support for the society is one of the reasons why religious writers and publications paid tribute to him following his death, from the evangelical *Record* to the society's own journal. In 1884, *The Spectator* described his attitude to the society 'as emphatic an answer to the detractors of missions as can well be imagined'. But this 'answer' does not indicate the complete change of heart that Sulivan suggested. Whatever his own religious beliefs, Darwin had always praised missionaries as a civilizing influence on 'savage' people, but he saw this influence in terms of culture, not religion. Though he eventually came to accept—with humility and pleasure—that even the Fuegian 'savages' were capable of civilization, there is nothing to suggest he ever revised his damning assessment of their original 'wild' state.

I discussed Darwin with the Yagán-heritage educator and writer Victor Vargas Filgueira at the annexe to the Museum of the End of World in Ushuaia, where he works as a guide. Filgueira is also the author of *Mi sangre Yagán* (My Yagán blood, 2022), a tender celebration of his great-grandfather Asenewensis, a shaman and one of the last generation of Yagáns to live according to their old customs and traditions. Filgueira's semi-fictionalized evocation of his great-grandfather's lost world is a moving counterpoint to the reductionist depictions of the Yagán people, and Filgueira has become something of an ambassador for his 'extinct' ancestors. He speaks regularly in schools, museums and universities, and he also gives an hour-long talk each week at the Museum of the End of the World about the culture and history of the Yagán. There were only five people present when I heard him, but his sense of purpose and urgency shone through his talk as he spoke without notes about the culture and history of the 'southernmost people in the world' and the misconceptions that had so often accompanied their interactions with Europeans.

Afterwards, I spoke to him about his childhood experience of racism; about the shame he once felt in school about his Yagán ancestry and the 'conflicting feelings, many contradictions' that led him to ask questions about his cultural identity and ultimately to embrace it. Like so many descendants of Argentina's First Peoples, his mother and Yagán-speaking grandmother were instrumental in this rediscovery. 'Our grandmothers were giants,' he said proudly. When I asked him why, he replied with a smile: 'Because they had great wisdom.' For Filgueira, the women of his grandmother's generation were the curators who kept 'the drama of our ancestry' alive, at a time when Yagán culture and traditions were still subject to the 'generalized non-existence' that followed Tierra del Fuego's colonization and its absorption into Argentinian and Chilean society. In his opinion, the history of the Yagán people is the familiar history of 'the Other' that has been repeated in many other colonial contexts.

I asked him whether the Yagáns still remembered Darwin. 'No. They aren't researchers,' he replied:

> They don't have the resources that I have. Nowadays, if I saw someone without clothes I would want to put clothes on them, so we can't blame people for the innocence of not knowing. The first impression is always going to be the main impression. Darwin's visual impression was 'these poor people, they're cold, they're hungry, they're skinny and tiny'. That was his conclusion, based on the most minimal first impression, which I might have reached, if I put myself in his place. The adjectives that were used afterwards is another question, but that first impression.

The 'innocence of not knowing' seemed to me to be a very generous interpretation of Darwin's responses to the Yagán, which obviated the prejudices and assumptions he and others had brought with them. Filgueira admitted that the 'prejudices of that time were terrible', but he was cautiously optimistic about his own times and encouraged by the greater willingness of Argentinians to appreciate Argentina's First Peoples. At the same time, he believes that true equality will not be possible until 'people look back and realize what happened to the Other in the place where they live'.

* * *

The colonization of the Yagán, Haush and Kawésqar 'canoe Indians' was not a violent process in the sense of military conquest. Nor was it a settler-colonial conflict over land and resources, but violence was not entirely absent from the colonization process. Interactions between European and Yagáns had been occasionally violent ever since their first encounters in the sixteenth century, and the intensity and frequency of these incidents increased towards the end of the nineteenth. In 1878, the crew of the salvage ship *Rescue* opened fire on a group of Yagáns in canoes, killing nine people. In 1881, a whaling crew kidnapped several Yagán women, killing fifteen members of their tribe who tried to stop them. Whalers and seal hunters frequently kidnapped Yagán women and sometimes forced Yagán men to work on their ships.

Nellie Burleigh described the 'wicked men ... some of whom were English' who forced sixty Yagáns to take refuge at the Bayly Island mission after a Yagán man had been shot and another given poisonous liquor. On another occasion, the Burleighs found an old man 'lying almost in the fire, and in a pool of blood' in the shelter when he tried to prevent miners from kidnapping his wife and daughter. In December 1893, the Burleighs took all the local women into the mission house to protect them from miners who 'sent a message to us that they intended either to shoot us or burn us out'. Following her return to England, Nellie criticized the traders, hunters and miners 'who ought to have known better—and who ought to have set the poor savages a better example', who 'bring discredit on Christian work, and who taught our people to swear and fight'.[26]

By the end of the century, such encounters had become so common that the Yagáns routinely hid their women and children at the approach of a ship. Nor were these fears limited to sailors. In the early 1880s, alluvial gold deposits were discovered on the coasts of Tierra del Fuego and the Strait of Magellan. Miners from Europe, the United States and South America converged on Tierra del Fuego, all attracted by a 'Patagonian Gold Rush' that rarely lived up to its expectations. Many of them were the 'blood-and-thunder fellows ... the very dregs of mankind, the froth of wickedness' described by Rockwell Kent in 1924. In the last decades of the century, there were sporadic clashes between miners and the diminishing groups of Yagáns who still lived on the beaches and coasts of Tierra del Fuego. And in the same period, miners also began crossing over to the northern shores of the Isla Grande from mainland Patagonia, where they found themselves in conflict with one of the most mysterious and little known of Tierra del Fuego's Indigenous Peoples.

18

THE GENOCIDE OF THE SELK'NAM

Until the late nineteenth century, the Selk'nam or Ona people ('people of the north') had had little contact with Europeans. They lived in the interior of the Isla Grande that they called Karukinká—the lands of the Selk'nam—in small nomadic bands, hunting the guanaco herds that provided them with their main source of food, clothing and shelter. Unlike the 'horse Indians' on the Patagonian mainland, the Selk'nam hunted the guanaco on foot with bows and arrows, using dogs to chase down their prey. Though the Selk'nam sometimes came to the sea in search of seals, otters or beached whales, they rarely encountered Europeans, and their contacts with the 'canoe Indians' who lived on the coasts and beaches were also infrequent. The Argentinian botanist and engineer Carlos Gallardo (1855–1938) described the Selk'nam as 'graceful types, well-formed, with small hands and feet and shapes that we may call aristocratic'. Both men and women, Gallardo wrote, were 'strong, agile, hardy, tireless, sober, demonstrating in all their physical exercises the excellence of these qualities'.[1]

This is how they appear in the photographs and films taken by the Salesian priest Father Alberto Maria de Agostini (1882–1960) in the early twentieth century, which show them wrapped in their knee-length guanaco capes with the fur turned outwards and guanaco-skin hats and head-dresses.[2] The goldmining boom effectively brought an end to centuries of isolation, as miners arrived on the beaches and shores of the Isla Grande that the Selk'nam regarded as their ancestral lands. To the Selk'nam, the miners were intruders and competitors for the guanaco herds. The goldminers regarded the Selk'nam as barbaric savages, to whom anything could be done with impunity. Selk'nam women were routinely kidnapped and kept as sexual slaves until they were visibly pregnant, whereupon they were allowed to return to their communities. The English adventurer James Radburne, who lived and worked in Tierra del Fuego as a young man in the late nineteenth century, described what appears to have been routine practice in the mining camps, in which even the 'crude and ugly women who were unimaginably dirty, so dirty that even these hard guys touched them as little as possible until they were washed', were tied up outside tents 'until they had time to take them to a river, there they cut their long hair and gave them a good brush with a small brush that each member of the gang carried for that purpose'.[3]

Some miners shot Indians on sight, while the Selk'nam responded to these incursions and the kidnapping of their women by ambushing miners, sailors and explorers, some of whom were tortured and mutilated. Initially, such confrontations were mostly limited to the coasts of the Isla Grande. But in the late 1880s, emissaries of civilization began to penetrate the interior of the island for the first time, and their encounters with the Selk'nam were often equally violent.

* * *

In November 1886, the Argentinian scientist, explorer and soldier Captain Ramón Lista (1856–97) arrived in Tierra del Fuego with twenty-five soldiers to conduct an exploratory expedition to the Isla Grande. Though Lista was sympathetic to the Tehuelche, his views of the Yagán and Kawésqar peoples were less positive. Citing Darwin, he described the Yagáns as a 'degraded race, that surely occupies the lowest level amongst all savage peoples' and as 'more savage than the Patagonians'—a difference he attributed to the isolation and insularity that had supposedly stunted their intelligence in the same way that 'any muscle atrophies that is not exercised'. If the Yagán people were 'degraded' Tehuelche, the 'dolichocephalic' skulls of the Selk'nam suggested the possibility that they might be the descendants of an 'invader race' from beyond the American continent. Lista tried to take anthropometric measurements of the

Selk'nam to explore this hypothesis, only to find that 'no savage lent themselves to my desires'.[4]

In his expedition journal, Lista describes how he and his men set out from Bahía San Julián in pursuit of a group of Selk'nam who set fire to their own encampment on their approach. On 25 November, Lista's men found another camp that had been set on fire and marched in search of its inhabitants. They soon found a large group of male Selk'nam, lined up in semi-circular battle formation, who showered them with arrows. Lista ordered his men to fire warning shots, but when the Selk'nam continued to shoot their arrows from a nearby thicket, he gave the order to open fire, followed by a sabre charge. By nightfall, twenty-eight Selk'nam were dead, while Lista's men suffered only minor injuries. Nine women and children were taken prisoner, given Spanish names and taken to Carmen de Patagones and Buenos Aires to be brought up as servants.

That same year, a Romanian mining engineer named Julius Popper (1857–1893) travelled through the Isla Grande on behalf of the Argentinian government. Popper is one of many Europeans to whom the sobriquet 'King of Patagonia' has been applied. Spears described him as

> an engineer of rare attainments—a civil, mechanical, and mining engineer—good in all three branches; an astronomer; a linguist who spoke and wrote a dozen languages fluently. He could with equal grace and precision conduct a lady to dinner or knock all the fight out of a claim jumper.[5]

Born into a Jewish family in Bucharest, and educated in Romania and Paris, Popper travelled to Asia, the United States and Latin America before arriving in Buenos Aires in 1885. In May the following year, he was named inspector of the goldmining companies in the Argentinian south. In September, he travelled to Punta Arenas and set out from Bahía Porvenir across the Isla Grande accompanied by seventeen armed men on horseback. After establishing a claim at the Bahía San Sebastián, he returned to the capital, where he published an album of eighty-four sepia photographs of the expedition in order to attract investment for his mining venture. This album included various photographs of Popper's men wearing their Austro-Hungarian-style uniforms at key moments in the expedition. In one photograph, a naked Selk'nam man holding a bow and arrows is lying dead alongside a half-destroyed wooden shelter behind three uniformed men, two of whom are firing their rifles into the distance. Another picture captioned 'Dead on the field of honour' shows two of Popper's expeditionaries firing from the other side of what seems to be the same body, without the bow and arrows.

The photographs were clearly staged, even though the corpse was real, and they were among the illustrations in the short book that Popper presented to the Argentine Geographical Society on 5 March 1887. Popper's text was sprinkled with unflattering ethnological observations of the Selk'nam people and descriptions of the brutal struggle that was unfolding on the island, where Indians 'continue killing sheep and horses whenever they can' while the guards or *puesteros* 'killed Indians whenever they caught them'. Popper blamed this confrontation on the 'alarming communistic tendencies' of the Selk'nam, who stole horses and sheep in the belief they were common property. Though he insisted that his men tried 'to enter into friendly relations with the Indians I met on the way', he also described an incident in which his expedition was attacked by 'about eighty Indians, with their faces painted red, and in a state of complete nakedness'.[6]

According to Popper, his men dismounted and fired at the Selk'nam with their Winchester rifles. This 'strange combat', as Popper called it, may well have been the subject of his macabre photographs, and it was not the only incident of this kind. Following this expedition, Popper was appointed technical manager of the Companía Anónima Lavadores de Oro del Sud (Southern Gold Washing Company Limited) and led two further expeditions to Tierra del Fuego, where he devised a fake 'cavalry' to frighten the Selk'nam—by mounting mannequins on horses made from hides, mud and grass. On 6 June 1893, he was found dead at his home in Buenos Aires, apparently from a heart attack, though his friends suggested he had been poisoned. Whatever the truth, there is little evidence that Popper played any significant role in the destruction of the Selk'nam people beyond his infamous photographs that provided a rare glimpse into the bloody confrontation unfolding on the island that only one side could win.

* * *

The municipal cemetery at Punta Arenas is one of the most hauntingly beautiful cemeteries in Patagonia. Rows of cypress trees line the rows of tombstones, brass doorways and chapels bearing names like Martinovic, Zlatar, Dacencic, Marasic, O'Reilly and Hamburger that testify to the multinational immigration that was part of the colonization of Magallanes in the nineteenth century. The cemetery's central square is dominated by an imposing mausoleum, with a cupola topped by a metal angel, adorned with sculptures and plaques from various societies that pay tribute to the man whose block-like bald head and handlebar moustache stares out from the commemoration plaques. This is the final resting place of José Menéndez Menéndez (1846–1918), one of the richest landowners in Patagonian history. Look closely, and you can still make out the faint splashes of red paint on the white stone, which suggest a less favourable form of remembrance that connects Menéndez to the Selk'nam people of Tierra del Fuego.

This history begins in October 1874, when the Argentine schooner *Rosales* docked at Punta Arenas in the midst of a scientific-military exploration of the Patagonian coastline between the Río Santa Cruz and the Strait of Magellan, whose members included the young Moreno and the Argentina-based German naturalist Friedrich Wilhelm Karl Berg. Among the schooner's other passengers was a twenty-eight-year-old Asturian immigrant to Buenos Aires named José Menéndez, who was then working as bookkeeper for two insurance companies in Buenos Aires. Menéndez's employers had sent him to Punta Arenas to settle the outstanding debts of a local Argentinian shipping captain named Luis Piedrabuena. By this time, Chile's southernmost settlement had made steady progress from the penal colony that Lieutenant Cambiaso's mutineers had all but destroyed in 1851. As O'Higgins had once predicted, the colony had become an important coal station and supply point on the Atlantic–Pacific steam navigation route. Punta Arenas had also become the centre of the *raquero*—from the English word for 'wreck'—salvage and rescue industry—in which local seamen stripped shipwrecked vessels in the Strait of Magellan and sold their contents.

Despite this limited progress, Punta Arenas remained a frontier town-cum-penitentiary where the garrison's soldiers were treated almost as badly as the convicts. The English settler William Greenwood described how military offenders in the colony were treated

> with the utmost severity, discipline was never relaxed, and punishment, principally by flogging, took place every day. For the smallest offence a man received a 100, 200, or even 300 blows with a 'Leña Dura' [hard wood] stick; if he could not stand it all the first day, he received the balance as soon as his lacerated back was able to bear it.[7]

In 1895, Spears stepped from the dock on to 'streets yellow with sand, then black with mud and glistening bright with pools of stagnant water', to find a city where '[c]owboys, shepherds, lumbermen, miners, and sailors gather ... to waste their substance in riotous living'.

Menéndez saw commercial possibilities in this unpromising outpost and bought Piedrabuena's assets in Punta Arenas in order to set himself up as a local trader. This may not have seemed the wisest decision in November 1877, when the garrison's naval gunners rose up against the hated governor Diego Dublé Almeyda. Though the governor escaped with his family, the garrison commander was killed and beheaded, along with other officers and civilians, before the mutineers and convicts burned most of the town to the ground for the second time in its history. Between 120 and 180 mutineers fled the city into the Patagonian interior, where most of them died or disappeared. The 'Gunners' Mutiny' proved to be a catalytic event for Punta Arenas and also for Menéndez. Following the revolt, the Chilean government abandoned the penal colony model and began to develop the Magallanes region for commercial sheep farming. The government indemnified all losses, including those of Menéndez himself. That same year, the Yorkshire-born landowner Henry Reynard established a sheep-ranch on Isla Isabel (Elizabeth Island) in the Strait of Magellan, with Cheviot sheep brought from the Falkland Islands, and its success encouraged others to lease land from the Chilean government for the same purpose.

Once again, Menéndez quickly recognized these opportunities and began renting land using a network of connections in the local and national Chilean government to evade the 30,000-hectare limit on land acquisitions. In 1884, he leased a 90,000-hectare ranch near Bahía San Gregorio, which became the foundation stone of a sheep-farming empire that spanned Chilean and Argentinian Tierra del Fuego and reached further north into Argentinian Patagonia. In 1893, Menéndez acquired Popper's 80,000-hectare grant in Tierra del Fuego following the Romanian explorer's death, which became the Estancia Primera Argentina, with 140,000 sheep. Menéndez collaborated closely with other entrepreneurs and landowners in the region, who lobbied, cajoled and bribed politicians in Santiago and Buenos Aires to ensure they could buy the lands they wanted and turn these leaseholds into permanent acquisitions.

These collaborators included the Portuguese *raquero* operator-turned-landowner José Nogueira, his Latvian wife Sara Braun and her brother Mauricio Braun, and Peter McClelland, the British representative in Chile of the trading firm Duncan, Fox & Co. In 1890, Nogueira obtained a twenty-year leasehold of 1 million hectares (over 3,800 square miles) from the Chilean government on the Isla Grade for sheep farming. Nogueira also co-founded the Sociedad Explotadora de Tierra del Fuego (Tierra del Fuego Exploitation Society) in order to raise capital for the expansion of sheep and cattle ranching. When Nogueira died before the society was legally constituted, his work was continued by Sara and Mauricio Braun and Menéndez, who became its principal shareholders. Under their leadership, the society became the driving organizational force behind the sheep and cattle industry in Tierra del Fuego and Patagonia.

In 1895, Mauricio married Menéndez's daughter Josefina, and in 1908 Menéndez and his son-in-law merged their companies into the Sociedad Anónima Importadora y Exportadora de la Patagonia (SAIEP), which combined ranching with slaughter, warehouses and shipping. Today, SAIEP is one of Argentina's largest supermarket chains, and one of many assets that have made the Menéndez–Braun family one of the richest dynasties in South America, whose wealth can be traced back to the 'empire of sheep' established by Menéndez in the desolate plains alongside the Strait of Magellan and the ancient lands of the Selk'nam.[8]

* * *

The sheep-farming industry posed even more of a mortal threat to the Selk'nam people than the mining industry. In fencing off their ranches and sheep stations, the landowners cut off the Selk'nam hunting routes and drove the guanaco herds southward and turned the Selk'nam into outsiders in their own lands. Fences along the coast also cut off their access to the sea and deprived the Selk'nam of another source of food. As Popper observed, the Selk'nam had no concept of private property, and they quickly came to regard the sheep they called 'white guanaco' as a consolation for the loss of their herds and an alternative source of food and clothing. In a letter to the legal advisor of the Sociedad Explotadora, Mauricio Braun complained that 'either the territory is left in the hands of the savages, or it is handed over to civilization'. Braun criticized the Chilean government for conceding 'great extensions of lands' to his fellow-ranchers, knowing that they were 'in the hands of the Indians', while doing nothing to prevent Selk'nam 'depredations'.[9]

In the absence of such action, Braun and his fellow-landowners tasked their estate managers, foremen and *campañistas* (shepherds on horseback) with killing, capturing and removing Indians from their estates. The Scottish shepherd William Blain, a sub-manager on a large sheep farm on the Isla Grande, later recalled how his estate manager gave him 'strict orders' 'not to allow any of the men to ill-treat the Indians, they were to be overcome by kindness'— orders he observed 'to the letter for a time'. Blain also described two fellow-workers nicknamed 'the Divel' and 'Buffalo Bill', whose 'sterner measures' against the Indians included killing the dogs that they used to run down sheep. In 1897, an anonymous New Zealand shepherd known only as G.H.C wrote to a Marlborough newspaper describing how estate workers on his farm hunted and captured male Indians known as 'Chunkies'—from the Selk'nam word *chonkóiuka*, which referred to the northernmost Selk'nam bands—and the women they called 'Chenas'—a corruption of the word 'China' that settlers used to describe Indigenous women on the Patagonian mainland. One of these Indian hunters was a man named Sam Hyslop who made horse hobbles from the 'skin of a *Chunkey* that he killed'. During his first Indian hunt, G.H.C was told to kill any Indian who shot an arrow.[10]

Participants in these expeditions were rumoured to receive a pound from the landowners in exchange for physical evidence of a kill, such as heads, ears or testicles. Greenwood recounts an undated incident that 'does not redound to the credit of the pioneers of civilisation, or hold out a lesson of charity and mercy to the savage' in which a party of armed white settlers pursued a band of Selk'nam sheep robbers and killed all its male members before cutting off their ears. When one Selk'nam girl offered 'the most notorious and successful "ear hunter" two rats as a peace offering, the ruffian drew his knife, made a show of receiving her present, and then deliberately struck her! Never mind, this was nothing!— one human being the less, perhaps a few sheep the more!'[11] This 'ear hunter' may have been the Scottish shepherd Alexander MacLennan, a veteran of Kitchener's army in the relief of Khartoum and the manager of Menéndez's first ranch in Tierra del Fuego, whom the Selk'nam called 'the red pig' because of his red hair.

MacLennan appears in Lucas Bridges' memoir under the pseudonym 'McInch'. Bridges described the 'uncrowned king of Rio Grande' as 'a most curious mixture. He who took open pride in having persecuted and murdered the Indians—albeit for their own good—hated to see a sore-necked yoke-ox, or a horse being unduly spurred.'[12] In one incident at a beach near Santo Domingo, MacLennan reportedly invited some 300 Selk'nam to a banquet. After getting them drunk, he ordered concealed marksmen to open fire on the gathering, killing dozens. In 1895, MacLennan and his men ambushed a group of Selk'nam families hunting seals and looking for molluscs on the shores of Cape Peñas, in which fourteen Selk'nam were killed. According to Bridges, MacLennan regarded this massacre as

> a most humanitarian act, if one had had the guts to do it. He argued that these people could never live their lives alongside the white man; that it was cruel to keep them in captivity at a mission

where they would pine away miserably or die from some imported illness; and that the sooner they were exterminated the better.[13]

* * *

The Selk'nam resisted these assaults with their limited means. They cut the fences of sheep farms, dug trenches to slow down their pursuers' horses and ambushed miners and settlers. Between 1890 and 1894, a Selk'nam band under the command of a chief named Seriot, known as 'Capelo' or 'Cardinal' in Spanish, after the shape of his hat, waged a guerrilla war against police, settlers and peons across the Isla Grande in both Chilean and Argentinian Tierra del Fuego. In one incident, Seriot pretended to befriend a group of miners and led them into an ambush. On another occasion, a Selk'nam ambush came close to killing the hated 'red pig' himself with an arrow. But in this unequal struggle, there could only be one outcome. Missionaries were also drawn into the conflict. In 1883, the Salesian Order of Don Bosco was permitted to establish an apostolic pro-vicariate in the north and centre of Patagonia, and another in Magallanes, which extended to the Isla Grande del Fuego and the nearby islands. In 1887, under the pro-vicariate's first prefect, Monsignor Giuseppe Fagnano (1844–1916), the Salesians established the San Rafael mission on Dawson Island, some 54 miles south of Punta Arenas, where small numbers of mostly Kawésqar 'canoe Indians' were sent for religious instruction.

In 1895, Fagnano proposed to the Sociedad Explotadora that captured Selk'nam should be transported from Tierra del Fuego to Dawson Island in return for paying the Salesians 1 pound sterling for every Indian they received. The society's president Mauricio Braun accepted this arrangement as 'the cheapest way we can get rid of them, short of shooting them, which is rather objectionable',[14] and in August that year the first large shipment of 165 Selk'nam was transported by the society to Punta Arenas to await a ship that would take them to Dawson Island.

On 8 November 1895, an anonymous letter on the 'Fuegian Indians and the Salesians' was published in the Santiago newspaper *El Chileno*, which related the horrendous events that unfolded in the city following the first transport of Selk'nam to Dawson Island. Published over two days, the article described the massacres unfolding on the Isla Grande, where Selk'nam men and women were shot 'as if they were flocks of guanaco'.[15] The author praised the Salesians for the 'arduous and grandiose task of civilizing these tribes' on Dawson Island and accused the governor of Magallanes, Miguel Señoret, of handing over Selk'nam deportees at Punta Arenas to his friends to work in sawmills or in other unpaid manual tasks and presiding over auctions in which children were forcibly taken from their mothers and given to local citizens as domestic servants. Some of the governor's 'little slaves', the author claimed, were handed over to *casas de tolerancia* (brothels) in order to 'serve as instruments of the most repugnant perversions'.

'When they realized that their children were being taken away from them,' *El Chileno*'s anonymous correspondent wrote, 'the Indians abandoned their habitual serenity and docility, and emitting horrible cries, tried desperately to defend their little ones.' Children who resisted were dragged from their parents and taken to their masters' homes, where they were tied hand and foot, while their mothers and fathers roamed the city, calling out their names. The remaining Selk'nam were left to sleep on the beaches or freeze in shacks that were hastily erected on the governor's orders. Some were given meat by locals who wanted to see 'savages' eating it raw. At least fifteen Selk'nam died of cold and hunger. Others drank themselves into a stupor or lay on the beaches, 'stretched out in the snow, almost naked, idle, stupid, abandoned to their fate', crying out so loudly that they kept the locals from sleeping.

These descriptions generated such a furore that the Chilean government ordered the Civil Court of Magallanes to open an investigation into the *vejamenes* (abuses) perpetrated against the Selk'nam under a local judge named Waldo Seguel, which confirmed many of the allegations in the *El Chileno* articles. One witness claimed that Alexander Cameron, the New Zealander administrator of the Caleta Josefina, the Explotadora's first ranch in Tierra del Fuego, had been paid 10 pesos for every Indian he killed. Another claimed that his employees had massacred Indians and killed them with poisoned sheep. A sheep-farm manager named

John Farquhar 'King' MacRae was accused of having killed eighty Indians in a single incident.[16]

MacRae, Cameron and the other estate workers denied all the allegations made against them. No testimonies were taken from the Selk'nam themselves, and the investigation continued until 1897, when it was inexplicably suspended. It was not until 29 February 1904 that Judge Seguel delivered his final judgment, which found that '[t]he Indigenous People of Tierra del Fuego were living in a state of barbarism; that they had no law of territorial property, and that they were nomadic, feeding themselves through hunting and, mainly, with the sheep they came across.' Because such people lacked any 'provision for defining their legal status', Seguel ruled, there was no basis for the 'criminal prosecution of any person for the abuses that are said to have occurred against the Indigenous tribes that have lived in Tierra de Fuego and adjacent islands'. By this time, there were few people—either in Chile or Argentina—willing to look more closely at an industry that had brought wealth and prosperity to the 'empty' territories of the far south.

In 1915, the Explotadora built a giant cold storage plant-cum-slaughterhouse at Puerto Bories, just outside the town of Puerto Natales. Using British-built refrigeration equipment, the Bories Cold Storage Plant was able to slaughter 4,000 sheep a day, with storage capacity for some 12,000 carcasses. Every year, mutton, wool and fat from some 300,000 animals were shipped from the jetty outside the plant, mostly to European markets. This system effectively commodified both land and sheep as the essential components of an industrialized conveyor system that connected the distant grazing lands of Tierra del Fuego and Magallanes directly to European banks, kitchens and textile mills through flows of capital and transportation links. In this context, there was no room for the Selk'nam. And the 'acts of cruelty and brutality' that the Salesian priest Alberto de Agostini later denounced as the primary cause of the 'rapid extinction of an inoffensive and vigorous race' were not just a 'shameful stain on civilization'—they were a direct consequence of the economic model that was constructed in the 'deserts' of the far south.

* * *

Before the introduction of sheep farming in the 1880s, there may have been 3,500 Selk'nam on the Isla Grande. In 1913, Lucas Bridges estimated there were only 300 members of this 'interesting tribe' remaining on the island. Some were killed by settlers or diseases or died in internecine conflicts with each other, as their lands diminished. Others fled to the south of the island to avoid deportation, where they found sanctuary at the Salesian mission in Río Grande, at the Anglican mission in Ushuaia or at Lucas Bridges' ranch at Viamonte. Some Selk'nam became peons on ranches and sheep stations and merged into the surrounding society. As was the case elsewhere in Patagonia, the imminent 'disappearance' of the Selk'nam and the other Indigenous Peoples of Tierra del Fuego increased their value to scientists concerned with the 'evolution of man' and racial difference.

Though Lista expressed his regret at the massacre of Selk'nam in 1886, his men spent the next few days stripping and combing scalps from the dead, and they also 'boiled and cleaned skulls and skeletons' under the direction of the expedition's surgeon and doctor, according to the governor of Argentinian Tierra del Fuego, Pedro Godoy. When the Selk'nam guerrilla leader/bandit Seriot was shot by police on the Bridges' ranch at Harberton in 1897, his skeleton was sent by Godoy to the Museo de La Plata, along with the remains of three other Selk'nam men. In 2016, all four skeletons were restituted to the Rafaela Ishton Selk'nam community.

In 1902, Lehmann-Nitsche visited Tierra del Fuego in his capacity as director of the La Plata Museum, where he took photographs and made physical measurements of twenty men and thirty Indigenous women at the Salesian mission in Río Grande in order to 'fix the characters' of peoples he believed were 'destined to disappear'. His subjects included a Yagán Indian woman, whom he defined as a 'relic of the most inferior ethnic groups', and two teenage Kawésqar girls in skirts and blouses, whom he found sitting outside the wooden 'wigwams' that had once been their family home. According to Lehmann-Nitsche, their parents had recently been shot by a white coypu hunter whose 'Winchester had triumphed over the bow and arrow of the unfortunate sons of the soil' and who now kept their daughters in the huts where he also stored his hides. One of the girls was photographed stripped to the waist, and her half-naked torso was later turned into one of the popular 'ethnographic postcards' of native women that blurred the distinctions between erotic and 'scientific' interest.[17]

Tierra del Fuego's vanishing peoples also found their way to Europe in the pseudo-scientific spectacles known as 'human zoos' that reached their zenith of popularity in the Belle Époque. In 1881, eleven Kawésqar were captured by a passing ship in the Strait of Magellan and taken to Paris, where they were presented to the public in a cage at the Jardin Zoologique d'Acclimatation in Paris as examples of the cannibals who 'eat their old women before eating their dogs' in times of hunger. Some 400,000 visitors came to see this spectacle in two months, including some of France's leading scientists. The ethnographer Julien Girard hoped that 'the most miserable and abject savages that we have ever had occasion to know' might provide insights into 'the first steps of humanity'. Topinard hailed the opportunity to study 'the last witnesses of an endangered age' without having to travel the world in search of them. The distinguished anthropologist and anatomist Léonce Manouvrier visited the exhibition five times on behalf of the Société d'Anthropologie and reported that '[w]e were able to take fifty measurements of each ... The only thing we could not do was to examine and measure the genital organs. It was not possible to see any lower than the upper part of the pubis.'[18]

The Kawésqar were taken to Munich, Hamburg and Berlin, where they were examined by the anatomists Hans Virchow and Theodor von Bischoff, who concluded on the basis of an examination of the younger females that the 'clitoris was small and not large as in the monkeys'. In Zurich, seven members of the group died from measles, tuberculosis and, in one case, from syphilis. The Swiss doctor Johannes Seitz treated these dying 'Fuegians' he called survivors from 'the cradle of humanity' and the 'remote bestial past'. Seitz also preserved their brains in formaldehyde and concluded that 'these savages are no different from Europeans from the cerebral perspective.'[19] Five members of the group were eventually returned to their homeland, of whom only four survived the journey.

Other 'Fuegians' followed the same trajectory. In 1888, eleven Selk'nam men and women were brought in chains on the steamship *Toulouse* to Europe by a Belgian whaler and impresario named Maurice Maitre, who set out to exhibit them at the Paris Exposition Universelle. Two of Maitre's captives died before reaching the French capital, and the remainder were exhibited in a cage near the Eiffel Tower, among the living exhibits of 'native peoples' from around the world. Maitre proudly presented the Selk'nam as cannibals and entertained the public by entering their cage with a whip as he fed them pieces of raw meat and fish. The Selk'nam were then taken to London, where the *Pall Mall Gazette* described their 'savage customs' as the 'talk of the town'. When one of Maitre's female exhibits fell ill and died, a doctor reported the case to the South American Mission Society, which denounced their treatment to the British government. To avoid arrest, Maitre fled with the Selk'nam to Brussels, where on 6 February 1890 the *Journal de Bruxelles* reported that a 'troupe of cannibals' had been arrested by the Belgian police on immigration offences. By the time the surviving Selk'nam were sent back to Tierra del Fuego, there were only four of them left alive. One died on the ship transporting them to Punta Arenas, and the remaining three died soon after their arrival on Dawson Island.

* * *

Today, Punta Arenas is a thriving city of just over 129,000, with a free port and a naval, air and army base, which also acts as a transportation hub for wool, mutton, propane gas, oil and frozen fish. Among the pretty blue- and red-roofed houses, the chic restaurants and cafés, the palaces built by Sara and Mauricio Braun in Punta Arenas have become national monuments, where visitors can admire the sumptuous French furnishings, Italian marble floors, billiard and musical performance rooms and wood-panelled walls that testify to the grandeur of the Braun–Menéndez dynasty and its empire of sheep. On the Avenida Bulnes, a bronze sculpture shows the famous shepherd Abel de Jesús Oyarzún Córdova leaning into the Patagonian wind with his horse and dog and a flock of sheep—the embodiment of the dogged pioneers who brought civilization to the 'wilderness' of the far south. On the promenade overlooking the sea, an elaborate allegorical sculpture of Captain 'Juan Guillermos' (John Williams)—the founder of Fort Bulnes—

and his crew stands near the statue of the Chilean naval officer Luis Alberto Pardo Villalón—the rescuer of Shackleton's expedition in 1916—pay tribute to Punta Arenas's history as a maritime city and the stepping-stone to Chile's Antarctic claims. In the Benjamín Muñoz Gamero Square, a statue of Magellan stands with his foot on a cannon staring towards the strait that bears his name, while two Indigenous figures—a Selk'nam and an Aónikenk—sit on either side of him with their weapons resting on their laps.

At the Maggiorino Borgatello Salesian Museum in Punta Arenas, a detailed ethnographic section on the culture and cosmology of the Selk'nam and other Indigenous Peoples from Magallanes and Tierra del Fuego is dramatized by 'heritage displays' containing life-size models of Kawésqar hunters spearing fish, Selk'nam shamans and traders. Unusually, the museum acknowledges the 'sickness, dispossession of lands, violence and death' that decimated Selk'nam society in the late nineteenth century while also praising the Salesians' 'humanizing' impact on the conflict. One display shows a furnished room containing a Chilean flag and a loom, where two Indigenous women wearing civilized clothes weave and sew under the benevolent gaze of a priest and nun. A letter from a child at the Dawson Island mission to the 'big father' Monsignor Fagnano in 1918 gives thanks for the life at the mission and promises 'to always be good. Always work hard to be your comfort in the earth and someday fly with you up there in the sky like birds, where they live with God, the Virgin and the Saints Forever.'

These depictions of the Salesians are not entirely false. Salesian priests privately protested the violence inflicted on the Selk'nam, and their missions were a preferable alternative to massacre and enslavement. Monsignor Fagnano later recalled how 'the poor Indians after having been stripped of the lands inherited from their ancestors, their houses violated and their animals stolen, are treated by the representatives of civilization with the most iniquitous barbarism imaginable'. On one occasion, Fagnano wrote to the Argentinian minister of foreign affairs accusing Menéndez of instigating his peons and the local police to 'hunt the Indians' on his estates, to which Menéndez responded with an open letter accusing the Salesians of harbouring 'thieves and murderers' at their La Candelaria mission in Río Grande.

These missions were not entirely benevolent institutions. At the San Rafael mission, the Salesians' own figures reveal a death toll from smallpox, measles and pulmonary infections among the Selk'nam and other Indigenous people who were sent there that was partly due to the cramped conditions in which they lived. One priest estimated that 40 per cent of the deportees died between 1896 and 1910. In 1896, a sea captain named Domingo Canales sent a report to the governor of Magallanes describing his 'horror and sadness' at the 'abandonment, the repugnant filth, the upsetting nakedness and misery in which lie some one hundred adult women and twenty or more men' that he observed on the island. According to Canales, the deportees became ill 'from the first moment of their contact with civilized races' and were left without medical attention or beds.[20]

Lucas Bridges met an old Selk'nam friend named Hektliohlh on Dawson Island who told him '[l]onging is killing me' for his lost lands. When Hektliohlh died soon afterwards, Bridges observed that '[l]iberty is dear to white men; to untamed wanderers of the wild it is an absolute necessity'. Though the Salesian missionary Father José María Beauvoir denounced the 'atrocious persecution' to which the Selk'nam were subjected at the hands of 'civilized white men', he attributed 'the germ of their dissolution' to their 'own physical constitution'. Unable to prevent 'the fatal blow that would finish off their bodies', his order had concentrated on saving 'the better part of them, in other words their spirit, redeeming their last remains, by the grace of Baptism'.[21]

* * *

For much of the twentieth century, this grim history received little public attention in Chile. In his *Pequeña historia fuegina* (Little Fuegian history, 1939), the historian Armando Braun Menéndez rejected what he called the 'calumny of persecution' surrounding the Selk'nam. The son of Mauricio Braun and Josefina Menéndez dismissed such accusations—and their implications for his own family—as isolated acts of adventurers and settlers and other 'police actions, common to our civilised centres'. Like his father, and so many of his contemporaries, Braun Menéndez attributed 'the principal cause of the extinction of the Fuegian aboriginal races' to 'their absolute lack of physical adaptation to the civilized life' and the fact that 'the step from the barbarism in which they were living to the civilization that was imposed upon them was too brusque'.[22]

Today, the ghosts of Patagonia's vanished Indigenous peoples haunt the backpacker destinations of Puerto Montt, Punta Arenas and Ushuaia. Murals depict the history of the Indigenous Peoples of Tierra del Fuego. Souvenir shops sell Selk'nam key rings, pillowcases and T-shirts, coasters and socks. Cut-outs of masked spirits from the Yagán Kina ceremony invite visitors into restaurants. On a bitterly cold morning, I took the *Patagones* ferry from Punta Arenas to Porvenir, on the north-west coast of Isla Grande, across the Strait of Magellan in search of a different kind of commemoration. Bahía Porvenir, or Future Bay, was once the centre of goldmining operations on the Isla Grande, and it was also the main point of contact between Punta Arenas and the sheep-farmers who colonized the island.

Porvenir is now a Chilean military base, with a population of just under 6,000, many of whom are soldiers. Some of them were on the ferry, staring sleepily at the television screens showing Adam Wingard's *Godzilla vs Kong* (2021). Most passengers were looking at their mobile phones, oblivious to the raucous din of explosions, growling and collapsing buildings emanating from the television. H.G. Wells—the father of so many science fiction catastrophes— once reminded readers of *The War of the Worlds* (1898) 'what ruthless and utter destruction our own species has wrought, not only upon animals, such as the vanished bison and the dodo, but upon its inferior races'. Like Darwin before him, Wells recalled how 'the Tasmanians, in spite of their human likeness, were entirely swept out of existence in a war of extermination waged by European immigrants, in the space of fifty years'.[23] In our new century of climate change-driven disasters, epidemics and mass species extinction, it is easy to forget the wars of extermination that were once waged against Indigenous Peoples, to which the annihilation of the Selk'nam belongs.

On the main road from the little harbour to the town of Porvenir, a large statue of a Selk'nam hunter stands like a sentinel in a guanaco-skin coat and holding a bow and a quiver of arrows.

This statue is the work of Richard Yasic Israel, a fifty-nine-year-old artist of Croatian-German heritage, whose great-grandfather came to Porvenir in the late nineteenth century, along with so many of his countrymen, to look for gold. Yasic's father was a mechanic, and his son is a furniture maker and self-taught sculptor and artist. The statue of the Selk'nam hunter was his first commission in 1999. At that time, the local council was unwilling to commission a sculpture from an unknown artist. Now his sculptures can be found all around Porvenir. The centre of the town has been re-named Plaza Selk'nam, and it contains a row of life-sized figures copied from the anthropologist Charles Wellington Furlong's (1874–1967) famous photograph of a Selk'nam band walking along a beach. A wooden plaque explains in Spanish and English the familiar euphemistic narrative of a 'tall, good-looking, strong and vigorous race', which 'succumbed to the impact of the colonizing action and the hand of the white man. Causing its rapid extinction.'

Yasic also showed me his recent mural, which stands just behind the Porvenir museum, entitled 'The Genocide of the Selk'nam'. It's a visceral, uncompromising work, based around the central motif of a crucified Selk'nam on a cross made out of sheep and dead and dying men, women and children, topped by Popper's minted gold coin. Around this central image are real episodes and characters from the Selk'nam genocide. There are Popper and his soldiers, standing next to an Indian corpse in the same posture from his infamous photograph. There is MacLennan, the red-headed Scottish murderer, receiving his payment from José Menendez next to a pile of Selk'nam heads. There are Salesian missionaries handing out the clothes carrying the germs that will take the Selk'nam to the nearby graveyard and a pile of heads gathered up in burlap sacks. It is not a painting for the Patagonian tourist industry. 'I felt it right here,' Yasic says, touching his heart with his fist. 'This kind of work wasn't always possible. When I was at school you never even heard what happened to the Indigenous Peoples of Southern Patagonia and Tierra del Fuego.'

Now there is more awareness. The 2008 report by Chile's Comisión Verdad Histórico y Nuevo Trato con los Pueblos Indígenas (Commission on Historical Truth and New Deal with Indigenous Peoples) described the fate of the Aónikenk, the

Selk'nam, the Kawésqar and Yagán peoples as a 'great tragedy. The greatest committed against the Indigenous peoples in Chilean territory … It was a process of extermination that took place. It was a genocide.'[24] In 2010, the remains of the five Kawésqar who died during their forced 'exhibition' in Europe were returned to Santiago from the University of Zurich and then transported to Punta Arenas for burial. Receiving these remains on behalf of the Chilean government, President Michelle Bachelet condemned the complicity of the authorities in exhibitions designed 'to satisfy the curiosity of the public and the anthropological interest of scientific circles', which she called the 'product of a racist attitude towards our First Peoples' and 'an affront to human dignity'.[25]

Yasic's mural is a testament to this recognition, and so is Felipe Gálvez's searing film *Los colonos* (The Settlers, 2023)—an equally unsparing account of colonial genocide at the tip of South America that echoes so many others. Yasic took me to see his fine wooden statues from the Selk'nam Hain or Klóketen initiation ceremony, overlooking a little bay away from the main town. A man was standing with his back to us, listening to 'Knocking on Heaven's Door' and drinking from a thermos as we stood shivering in the cold, beneath an electricity pylon and a row of antennae. Next to us stood the muscular painted body of a Selk'nam shaman with his fists raised looking towards the land from which his ancestors were 'exterminated'—not in the passive sense that Darwin meant it, but in its original active meaning of the Latin verb *exterminare*: 'to send beyond a boundary or frontier, expel, banish'—and also to kill.

The same figures also appear in the 'noche Selknam' (Night of the Selk'nam) that takes place every year in Porvenir, where musicians, actors and artists pay homage with fires, lights and theatrical performances to the First Peoples who once roamed the island. And as I stood looking over the barren hills beyond the town, with Dylan playing in the background, I imagined them like the figures in the photographs Yasic's park was taken from: resilient and determined, masters of the lands in which they had survived for thousands of years—unaware that the photographer who captured that image was recording a society that he believed would soon vanish from the earth.

Fig. 9: Exhibition room at the Museo de La Plata. By 1890, 113 standing skeletons and thousands of crania, bones, death masks and mummies enabled the visiting public to trace the evolution of humanity in the Americas from prehistory to the recently conquered 'desert'.

Fig. 10: Wives of Inacayal and Foyel, with other Indigenous prisoners. Photographer unknown, but probably taken at the Tigre barracks in 1884. Subsequently published as part of a series on 'aboriginal iconography' by M. A. Vignati, in the *Revista Museo de la Plata*, 1942.

Fig. 11: A Yagán family in 1883, taken by an unknown photographer during a French expedition.

Fig. 12: Selk'nam women, circa 1905. Photo by Alberto de Agostini.

Fig. 13: Group of 'Indian hunters', 1905. Photographer unknown, but taken on a ranch in Tierra del Fuego.

Fig. 14: A Kawésqar man, Tierra del Fuego, 1923.
Photo by M. G. Hentschel.

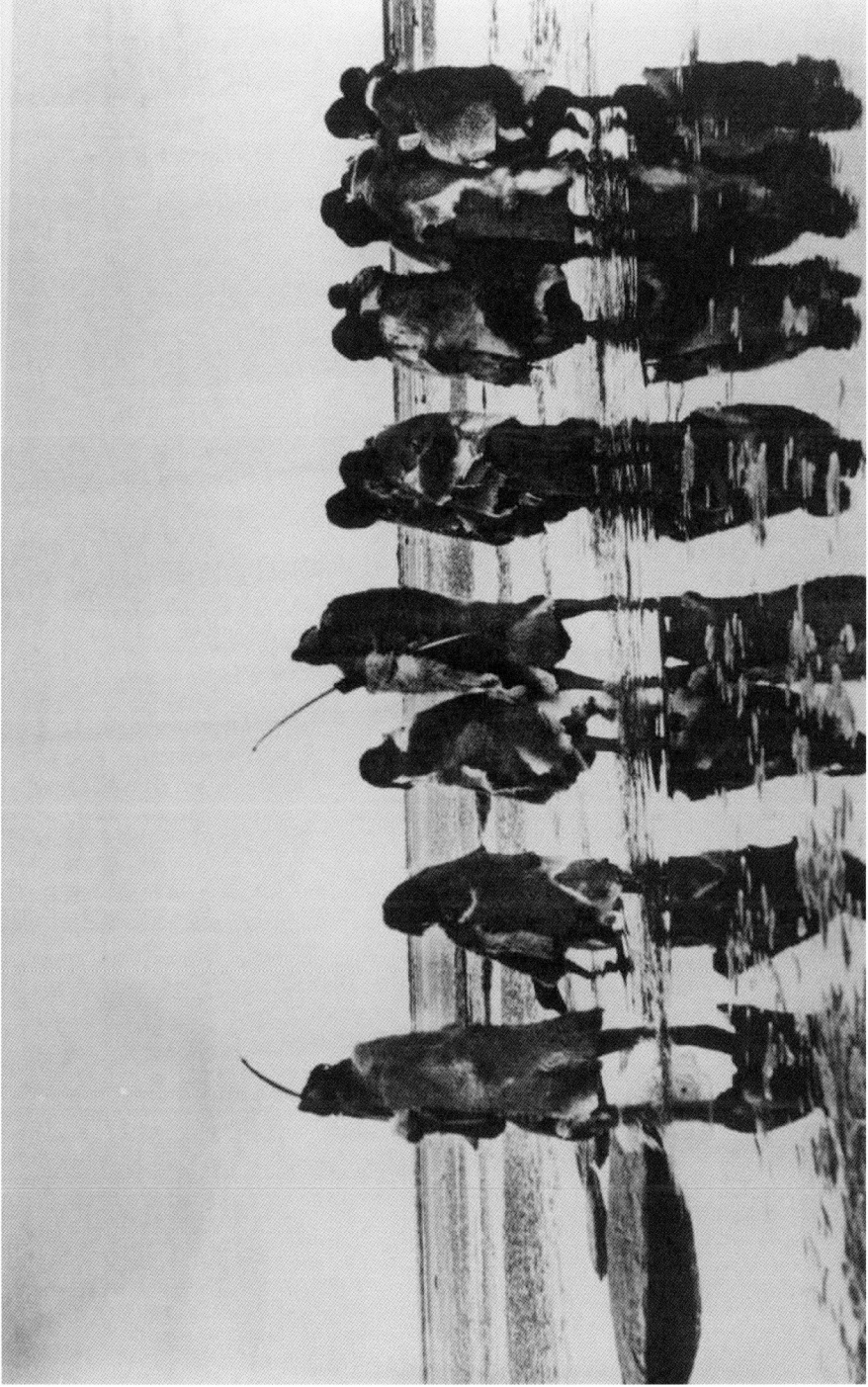

Fig. 15: Selk'nam family crossing the Río Fuego, Tierra del Fuego, 1908. Photo by Charles Wellington Furlong.

Fig. 16: *Patagonian Giants* by Jessie Tarbox Beals. The photo shows a group of Tehuelche 'exhibits' at the 1904 World's Fair in Louisiana.

19

THE WILD SOUTH

In 1910, Argentina invited the world to celebrate its centennial with a series of celebrations and exhibitions from which no expense was spared. Throughout the year, presidents and international dignitaries came to Buenos Aires to visit the pavilions celebrating Argentina's achievements in industry, agriculture, hygiene, railways and electricity, in addition to art exhibitions, operatic and theatrical productions, gala events, horse races and international scientific congresses. At that time, Buenos Aires was the largest city in Latin America and the capital of a country that seemed destined for greatness. Foreign banks, elegant hotels and Parisian boulevards proclaimed the wealth and elegance of a city that was connected to the outside world through shipping and railway companies that extended to Britain, the United States and Central Europe. The sterile plains that Darwin had observed around the mouth of the Río Negro in 1832 were cultivated with wheat and alfalfa; the Pampas were crisscrossed with giant farms and ranches enclosed by barbed-wire fences. Telegraph wires and some 35,000 miles of railway lines—more than any other South American country—reached deep into the interior, bringing agricultural products to the ports of Bahía Blanca or Buenos Aires and transporting travellers all the way from the Atlantic coast to the Andes in Pullman saloons and restaurant cars.

Italians, Spaniards, Russians, Poles, Germans, 'Turcos'—anyone from the Middle East—and a 40,000-strong British community that constituted the largest outside the empire—all these nationalities were part of a net annual immigration that averaged 100,000 and sometimes reached 200,000 per year by the 1890s.

The 1900 *Baedeker of the Argentine Republic* boasted that 'the basis of the population is European ... All the civilized races of the earth have met here.'[1] Between 1869 and 1895, the Indian percentage of the overall Argentine population fell from 5 per cent to 0.7 per cent. In Rosas's time, Afro-Argentines composed 25 per cent of the population of Buenos Aires. By 1887, they constituted 2 per cent. Few lamented these outcomes. Unlike the United States, boasted the 1900 *Baedeker*, Argentina's 'ethnical composition ... does not make us fear any complication between the races', because

> we do not know anything about the Indian, Negro or Chinese problems. The Indians who live amongst us have been converted to the Catholic religion and have adopted its humanitarian principles; many of them have been killed by consumption and the rest have been incorporated into society as a powerful working element.

The Afro-Argentine population had succumbed to 'that inflexible biological law which condemns the inferior organisms [to disappearance]'. Chinese immigration was also minimal, 'because the Constitution imposes on the Government the duty of favouring white immigration against yellow'. The result was a racial mixture in which 'the European blood has prevailed by its superiority, regenerating itself constantly by immigration'.

There were cracks in this glittering edifice, some of which anticipated Argentina's later political crises. Most of Argentina's immigrants in the late nineteenth and early twentieth centuries did not come from northern Europe, as Sarmiento and Alberdi had hoped, but from Italy and Spain. They brought with them political ideologies such as socialism and anarchism, which led a new generation of nationalist intellectuals and politicians to question the model of a Europeanized Argentina and consider whether these emissaries of civilization were as civilized as they should be. Where Argentinian scientists such as Holmberg, Moreno and Zeballos had once applied Darwinist and Social Darwinist ideas to the elimination of Argentina's 'inferior' races, their successors flirted with eugenicist ideas in the hope of eliminating the unwanted consequences of immigration.[2]

These anxieties were exacerbated by militant labour protests and strikes in both Argentina and Chile in the first two decades of the twentieth century, to which the authorities in both countries responded with police and paramilitary violence, draconian anti-union legislation and laws facilitating the deportation of 'undesirable' foreigners to their countries of origin. Between 1919 and 1922, the labour agitation in the cities reached rural Patagonia, where a wave of strikes spread through ranches in Chile and Argentina. These disputes were suppressed by the army and security forces in both countries. In Argentinian Patagonia, the army massacred an estimated 1,200 workers in 1921, whose bodies remain buried in secret mass graves on the ranches where the strikes took place. The tragic events of 'Patagonia rebelde' (Rebel Patagonia), were one of the few occasions in which Patagonia attracted the attention of the central government and the national press. Far from the glittering Argentinian capital, and the cities of the north, Argentina's southern territories remained a land apart. Militarily conquered, and economically integrated into the national and international economy, the territories that Darwin and other travellers had brought to the world's attention as an exotic and mysterious wilderness held little intrinsic interest except as grazing lands for wealthy farmers and absentee landowners.

* * *

Descriptions of Patagonia in the 1900 *Baedeker* read more like an investment prospectus than a travel guide, with lists of statistics on the size of ranches, most of which were foreign-owned, and details of the size of their herds and breeds of sheep or cattle. Though *Baedeker* looked forward to a time when tourists would come to visit the forests, lakes and mountains around Nahuel Huapi, its sketchy descriptions of the few ports and settlements were unlikely to attract any but the most intrepid traveller. The 1912 National Territories Census in Argentina lists only 95,486 inhabitants in the five Patagonian governorships, more than half of whom lived in the northernmost governorships of Río Negro and Neuquén.[3] Chilean Patagonia was equally under-populated: the 1906 census of Magallanes counted 13,309 inhabitants in the territory, of whom more than 10,000 lived in Punta Arenas.[4]

This sparse population was partly due to an agro-export business model, which required large tracts of grazing land for sheep and cattle but very few people. Despite the fortunes that could be made from large sheep stations or cattle ranches, only a few settlers had the capital to establish them or the determination and resilience to live in a territory that was still regarded as distant and inaccessible. In Argentina, most of Patagonia's population was concentrated on the Atlantic coast, in port-towns like Río Gallegos, Puerto Deseado, Ushuaia and San Julián that acted as pick-up points for shipments of wool and mutton. In 1898, the Buenos Aires Great Southern Railway Company built a railway station at Choele Choel that connected the former 'bandit capital' to a line of settlements from Bahía Blanca to Neuquén.

In 1908, this station was replaced by a new and larger nine-track station, named the Darwin Railway Station, which became the nucleus of a town of some 2,000 inhabitants, with a cinema and a hotel catering to travellers who took the 870-mile overnight train from Buenos Aires. But its main purpose was the transportation of fruit from the orchards that lined the Río Negro all the way to the former 'Country of Wild Apples' in the east, to Bahía Blanca and Buenos Aires. Further south, Tierra del Fuego was almost entirely given over to sheep-farming, along with much of Chilean Patagonia north of Punta Arenas.

Beyond these settlements, Patagonia consisted mostly of isolated farms and giant sheep and cattle ranches, many of which belonged to foreigners. In 1910, the American mining engineer Bailey Willis attended the International Scientific Congress as a delegate of the Smithsonian Institution along with the Czech anthropologist Dr Aleš Hrdlička, the Smithsonian Museum's first curator of physical anthropology. During the congress, Willis was invited by José María Ramos Mejía, the minister of public works, to conduct a topographical, geological and hydrological survey of northern Patagonia with a view to developing the region. In 1911, Willis's Comision de Estudios Hidrologicos (Commission of Hydrological Studies) began a two-year expedition financed by the Argentinian government, which attempted to establish a route for a railway line through northern Patagonia and realize Ramos Mejía's visionary scheme for an 'industrial city' on the shores of Lake Nahuel Huapi.

Willis's team also explored a proposal from Moreno to create Argentina's first national park from a government grant of 30,000 square miles of land (subsequently increased to 40,000) around Lake Nahuel Huapi in reward for his work on the Chile–Argentina boundary dispute.

Willis settled on Bariloche as the most suitable location for a railroad station that would bring tourists to the shores of Lake Nahuel Huapi. In this 'struggling village of shingled cabins', where the streets were 'dust in summer and mires in winter', his team surveyed potential hotel sites and roads that would one day bring tourists to villas and summer homes in the surrounding lakes. Willis also identified a suitable site for Ramos Mejía's 'industrial city' on the north-eastern banks of Nahuel Huapi, and his team drew up plans for a damn, a sewage and water system and a town with streets, monuments, neighbourhoods, a military quarter and a sports field. Though Willis was optimistic about Patagonia's long-term prospects, he was less positive about the immediate future of a region requiring 'law enforcement, order, and development', which was 'useless to expect ... from the miserable corrupt officials in power'.[5] The Willis Commission's proposals were accepted by Ramos Mejía but scuppered by a combination of bureaucratic inertia and political wrangling over how it would be financed. In 1937, Willis received a request from the Argentinian government to buy his maps of northern Patagonia. The following year, Moreno's park was established, which laid the basis for Bariloche's transformation into the 'Argentinian Switzerland' that would later attract Nazi criminals and well-heeled tourists from Buenos Aires.

* * *

This 'industrial city' never materialized, and it was always an optimistic aspiration in a region that still exhibited some of the more undesirable features of a frontier zone. Willis often observed the efforts of the poorly resourced *policia fronteriza*—frontier police—to contain the outlaws and bandits who infested Patagonia in the first decades of the twentieth century. Butch Cassidy and the Sundance Kid, the 'English bandit' Helen Greenhill and a succession of Chilean desperadoes all contributed to Patagonia's reputation as the 'Wild South' and a refuge for the outcasts of civilization. In 1900, the governor of Neuquén complained to his superiors that police were powerless to stop the well-armed 'hordes of bandits captained by Chileans' who were crossing the border to raid ranches close to the Andes, stealing animals and money and leaving a trail of homicides in their wake.

If Patagonia was a refuge for bandits, it was also a place where Argentina sent criminals and political prisoners. In 1902, the Argentinian government began the construction of a 380-cell *presidio* (penitentiary) in Ushuaia, much of which was built with convict labour. Like the Chilean governments of the nineteenth century, Argentina's rulers saw the remoteness of the far south as a 'natural prison' for social undesirables—some of whom they regarded as the unwanted consequences of European immigration—where the frigid temperatures of Tierra del Fuego were part of their punishment. Convicts were transported by ship in leg-irons from Buenos Aires. A former warden later described the arrival of these 'phantoms, spectres' in the 'white hell' of Ushuaia after a twenty-nine-day voyage, where they emerged from the hold '[s]hrunken, unshaven, their ankles raw from the chafing of their leg irons, their thighs scratched and bleeding, their clothing wrinkled like handkerchiefs or towels'. Convicts endured a harsh prison regime—even minor breaches of discipline were punished by forcing offenders to lie naked in the snow before being brought inside for an ice-cold bath.[6]

Escape was not a possibility. The few who tried were soon caught. Most were thwarted by the climate or the terrain and gave themselves up in order to avoid starvation or freezing to death. Others hanged themselves or died of TB and other diseases that were a routine consequence of incarceration in *el último rincón*—the furthest corner—of the republic. It was not until 1947 that Juan Perón finally closed the prison, which now caters to the 'dark tourism' market, where visitors can eat snacks in the penitentiary exercise yard or have themselves photographed in cells with life-sized models of child killers and political prisoners. But even with its bookshop, café and theme park accoutrements, the thick walls and cheerless, claustrophobic cells still exude a vague, oppressive horror of the prison at the end of the world where urban Argentina once warehoused dangerous criminals or political deviants, knowing that many of them would never return. And beyond these harsh outposts of civilization, and the farms and ranches that contributed so much to Argentina's prosperity, the vanquished Indigenous Peoples of the south lived a largely marginal and—from the perspective of the wider world—almost invisible existence in the lands where they had once roamed freely.

* * *

A poster at the Railways and Land Transport Exhibition in 1910 shows a startled Indian holding a lance, fleeing from a steam train. This was the generic image in urban Argentina of its conquered Indigenous Peoples—the barbarian-savage, recoiling from the forces of modernity and progress that also doom him. 'There are estancias within a hundred leagues of Buenos Aires which we remember as desert country in the power of the Indians, where now traps and carriages of English type are seen crossing the plains, where folk dine in in evening dress in luxurious homes,' wrote the authors of *The Argentine in the Twentieth Century* (1911).[7] In a 1910 handbook on Argentina, written for his countrymen to coincide with the centennial, the English writer William Alfred Hirst described Roca's 'reduction of Patagonia' as 'an event the magnitude of which cannot yet even be estimated', which had brought 'an enormous waste, a no-man's land, unmapped and roamed over by savage Indians' under the control of the state.[8]

These 'savage Indians' had not entirely disappeared, but Hirst was not the only writer to conclude that they had either vanished from history or were about to be expunged from it. In *Los Indios Tehuelche: Una raza que desaparece* (The Tehuelche Indians: A disappearing race, 1895), Lista—the scientist and explorer who had hosted the Tehuelche chief Orkeke and massacred twenty-nine Selk'nam ten years earlier—attempted to preserve the memory of what he called 'a race about to disappear from the world stage' with a glossary of Tehuelche words and retellings of Tehuelche myths. Unlike some of his scientific contemporaries, Lista rejected 'the law of natural evolutionism' as an explanation for the demise of the Tehuelche, which he attributed to gunpowder, alcohol and the 'limitless cruelty' and 'pillage' perpetrated by his countrymen. This combination, Lista argued, had brought about the 'downfall of an ancient, American race, that even if only out of scientific interest, let alone humanitarian sentiment, we should have protected and allowed little by little to merge into the civilized masses'.[9]

Other visitors were equally pessimistic about the future of the Tehuelche. The Swedish polar explorer Otto Nordenskjöld led various mineralogical expeditions to Patagonia in the 1890s, which led him to conclude that 'the day cannot be far off when the last Patagonian in the old sense shall have ceased to exist'. In 1900, the British explorer, soldier and cricketer Hesketh Vernon Prichard visited Patagonia in search of the 'prehistoric Mylodon'—an expedition financed by the *Daily Express* in exchange for a journalistic account of his travels. Prichard lamented the disappearance of the 'free and a happy life' the Tehuelche had previously enjoyed and their depleted population, which led him to regard himself as 'their last chronicler' before they were 'brushed off the face of the earth by the sweeping besom that deals so hardly with aboriginal races, and is known as "civilisation".'[10]

In 1904, the British army officer Colonel Thomas Holdich, a member of the British government's Arbitration Tribunal on the Chile–Argentine Boundary dispute, similarly described the Kawésqar in the Strait of Magellan as 'the last specimens of a disappearing race of humanity, but they were not alluring'. Holdich was no more enamoured with the Yagáns and Selk'nam he observed at the Bridges' farm at Harberton. 'Civilization has come upon them ere their development was equal to it,' he wrote, while the Fuegian peoples

> are not ill-natured, and they are teachable and make good servants, but these are only animal qualities; they are the last remnants of a world's race that in fullness of time must inevitably give place to others higher in the scale of human intelligence, unless indeed the great process of national development is to go backwards.[11]

Other writers saw Tierra del Fuego's Indigenous peoples in similar terms. In a visit to Patagonia in 1898, the Argentinian journalist Roberto Payró reported that 'the Fuegian is vanishing with shocking rapidity … They have been hunted like beasts, in the name of the highest principles of humanity.' The explanation for this disappearance, Payró argued, could be found in the writings of 'Darwin ... and many other anthropologists, [who] noticed that whenever the European passes an Indian dies and disappears, attacked by natural and artificial enemies that tend to remove him so one more fit can replace him.' In Tierra del Fuego, 'as in the Pampa and every other region inhabited by savages,' Payró wrote, 'the superior races have, in effect, supplanted the inferior, first destroying them ... The Indians of the extreme south of America cannot remain exempt from this general law, and they have not been.'[12] Payró's account of his Patagonian travels is sprinkled with references to Darwin's travels and Darwinian concepts. Observing a female English immigrant he called 'Miss Mary' flirting with an Argentinian passenger on his ship, he saw the couple as the embodiment of 'the new energy that will construct the powerful energies of work in Patagonia' through the union of Argentinian masculinity and the ideal 'Patagonian settler' and representative of 'this strong weaker sex [that] has already displaced the intelligent and strong indigenous Tehuelche women in Patagonia, a scarce specimen today'.[13]

History does not record whether Miss Mary embraced the 'evolutive task as feminine members of the masculine civilization that have risen with analogous climates', but statistics bear out Payró's predictions regarding Patagonia's Indigenous Peoples. The 1895 Argentinian National Census estimated the Indigenous population of Patagonia and Tierra del Fuego at 19,850 and 2,000 respectively, most of whom had not been censused. By 1912, the number of 'non-censused savages' in the Patagonian provinces of Santa Cruz and Chubut was estimated at 1,000 and 260 in Tierra del Fuego, compared with an overall figure of 11,540 'censused Indigenes' in Patagonia, the Pampas and the Chaco region. Though the authors noted that Argentina's Indigenous Peoples 'are not disappearing with the rapidity expected', the census figures demonstrate that they were 'being mixed, civilized, and diluted into the general mass of the population'. In Chile, the 1907 national census referred to 24,100 'pagans' left over from the 'numerous and hardened population of barbarian tribes' who had 'only been incorporated into civilization and duly colonized, in the last quarter of a century'.[14] This category referred exclusively to the Mapuche peoples who had been forcibly incorporated into reservations or *reducciones* following their military defeat in 1883 that brought the pacification of Araucanía to an end. In the southernmost province of Magallanes, no Indigenous People were recognized in the census. On paper, the First Peoples of Chilean and Argentinian Patagonia no longer existed, or else they were disappearing. And as was the case in many other countries, the 'vanishing Indian' became a recurring theme and an object of morbid fascination among the travellers and visitors who continued to follow Darwin's footsteps to the wild south.

20

SALVAGE AND SURVIVAL

In 1900, the British mathematician, eugenicist and white supremacist Karl Pearson gave a lecture on the subject of 'National life from the standpoint of science' to the Literary and Philosophical Society in Newcastle against the background of the Boer War. This was a classic example of the doctrine known as 'Social Darwinism'—the adaptation of evolutionary ideas to wider discussions about race and class, which depicted human history and society as the result of natural selection and the 'struggle for existence'. Like many similar examples, this quasi-Darwinian interpretation of history was the product of a very specific period of history. Pearson delivered his lecture at a time when Britain was struggling to defeat Boer guerrilla fighters in South Africa, and the British government was concerned at the physical deterioration of its armed forces. In this context, Pearson invited his audience to consider 'mankind as a product of Nature'. He argued that races and nations, like species, required 'conscious or unconscious selection' to 'reduce the tendency to the production of bad stock' in a world where civilization was the result of 'the struggle of race with race, and the survival of the physically and mentally fitter race'. If the 'continuous progress of mankind' was the product of the 'scarcely recognized outcome of the bitter struggle of race with race, being subject to the stern law of the survival of the fitter', then Pearson urged his audience to acknowledge that '[t]he path of progress is strewn with the wreck of nations; traces are everywhere to be seen in hecatombs of inferior races, and of victims who found not the narrow way to greater perfection'.[1]

Pearson's talk contained many of the essential tenets of 'Social Darwinism' that accompanied the high watermark of empire and colonialism: the application of the 'struggle for existence' and natural selection to races and nations; the inevitability of racial 'extinctions' as the collateral damage of civilization; and the belief that civilization was the product of white racial superiority. Only a few, largely marginal voices, challenged this consensus. The Russian anthropologist and zoologist Nicholai Nikolaevich Miklouho-Maclay (1846–88) conducted extensive fieldwork among the Indigenous Peoples of Papua New Guinea in the early 1870s, which led him to reject the idea that 'the dark races, as the more weak and lowly, had to disappear giving way to the white variety'. Miklouho-Maclay attributed the 'extermination' of such peoples to 'none other than the use of gross force ... Any honest man should rise against such atrocities.'[2] In *De l'égalité des races humaines* (The equality of the human races, 1895), the Haitian anthropologist Joseph Auguste Anténor Firmin (1850–1911) argued that '[h]uman beings everywhere are endowed with the same qualities and defects, without distinctions, based on color or anatomical shape'.[3] In *The Mind of Primitive Man* (1910), the German-American anthropologist Franz Boas (1858–1942) rejected the idea of a 'series of stages of culture which represent for the whole of humanity an historical sequence'—a classification that he argued was 'stimulated by the work of Darwin and his successors'. Boas believed that '[t]here is no fundamental difference in the ways of thinking of primitive and civilized man' and criticized the 'logical fallacy of first assuming that the European represents the highest racial type and then interpreting every deviation from the European type as a sign of lower mentality'.[4]

The calamitous events of twentieth-century history also led scientists to question the binary distinctions between the civilized and the savage. In a 1954 special issue of *The UNESCO Courier* dedicated to the 'last frontiers of civilization', the Swiss-born ethnographer Alfred Métraux urged world governments to correct the 'errors of the past' through which 'the crushing of primitive people in the path of our civilization was accepted as a natural state of affairs' and 'Darwinism served as a legitimate excuse for the extermination of weak, backward tribes.' The same issue contained an article by the French anthropologist Claude Lévi-Strauss, which dismissed the notion that '"primitive peoples" had no history, merely because they could not read or write' and warned scientists to avoid 'ambiguous terms like savage, primitive, or archaic' when referring to societies characterized by the 'presence or absence of some form of writing'.[5]

By this time, scientific racism had lost much of its intellectual force, and the 'ambiguous' categories that Lévi-Strauss described were no longer taken for granted. Lévi-Strauss's fieldwork among Brazilian First Peoples was part of a generalized shift in anthropology away from racial classifications and the hierarchies that accompanied them. In his introduction to *The Savage Strikes Back* (1937), the Polish-British anthropologist and ethnographer Bronisław Malinowski urged his fellow-anthropologists to 'break down the barriers of race and of cultural diversity' in order to 'find the human being in the savage'.[6] To some extent, this process was already underway, as anthropologists engaged in extended fieldwork with the peoples they studied. And in Patagonia, some of these anthropologists found their way to the territories where Darwin had once seen 'savages' in their 'natural state' for the first time.

* * *

In the first half of the twentieth century, there was a notable shift in the way Patagonia's First Peoples were portrayed, which coincided with a wider shift in the various disciplines concerned with the 'science of man'. The American artist, explorer and photographer Charles Furlong made various visits to Tierra del Fuego in the early twentieth century, using local Indigenous guides, in which he recorded songs and chants and took photographs of the Yagán and Selk'nam peoples. Furlong wrote about 'the Southernmost people of the world' in a series of articles for *Harper's Weekly* that were not immune to the dehumanizing language of his predecessors. In one of these articles, he described his discomfort on seeing his Selk'nam guide Aanikin's 'dark beady looking eyes' staring at him with a 'wolf-like expression which no white man likes to see lurking in the face or eyes of the savage or man of the wilds'.[7]

In response to a questionnaire from the American psychoanalyst Isador H. Coriat on 'Psychoneuroses among Primitive Tribes', Furlong revisited the same episode and compared his guide's attitude and expression to 'the look a dog takes on sometimes before he snaps'.[8] Despite these reductionist 'savage' stereotypes, Furlong attempted to approach the Selk'nam with 'an open mind and an open heart' and avoid 'prejudice and warped judgement'. He denounced the 'cruel and persistent warfare' waged by white settlers against the Selk'nam and the atrocities, deportations and diseases that had brought 'this splendidly picturesque tribe' to the brink of disappearance. Furlong called on the Chilean and Argentinian governments to show 'magnanimity' and give the survivors a grant of land before 'they follow those aborigines of another sea-girt land, Tasmania, where to-day not a Tasman is left'.

No such magnanimity was shown. In 1934, the writer and historian Ricardo Rojas was sent to the Ushuaia penitentiary for his opposition to the military dictatorship that overthrew President Hipólito Yrigoyen. As a high-ranking political prisoner and respected intellectual, Rojas enjoyed more freedom of movement than his fellow-prisoners, and he was a curious and sympathetic observer of the Fuegian Indians he encountered on the island during his four-year imprisonment. His book *Archipiélago Tierra del Fuego* (1947) was written, in part, to exorcise what he called 'the nefarious legend created a century ago by Darwin', which he believed had led his countrymen to neglect Argentina's southernmost territories and ignore their responsibilities towards their Indigenous inhabitants. Rojas criticized Darwin's ethnological errors and his damning descriptions of Tierra del Fuego as a 'miserable land, the most inhospitable in the world, inhabited only by almost bestial cannibals'. He celebrated Tierra del Fuego's 'strange beauty' and the culture, character and resilience of its peoples and described the 'abrupt disappearance of the Fuegian Indians' as a loss to science and the national economy, 'because in exterminating them we have not known how to replace them'. Rojas called for a more humane administration based on 'reason and justice' that would address the 'almost criminal dehumanization' of its peoples.[9]

Once again, these calls fell on deaf ears. Other visitors also criticized the dehumanization of Tierra del Fuego's First Peoples in similar terms. Much of what is now known about the customs and beliefs of the Selk'nam and Yagán peoples is due to the German priest and anthropologist Martin Gusinde (1886–1969). Between 1916 and 1917, Gusinde carried out fieldwork among the Mapuche for the National Museum of Ethnology and Anthropology in Santiago before making his first visit to Tierra del Fuego in 1918 to study the Selk'nam. Gusinde visited the Salesian Missions at Dawson Island and Río Grande and wrote of his 'deep grief and profound discouragement' on encountering Indigenous Peoples impoverished by 'the forcible acquisition and theft of their lands, invaded and occupied by the civilized, [which] removed their means of subsistence'. He denounced what he called a 'premeditated plan, put into practice with refinement and without any consideration whatsoever', which constituted the 'principal and true cause of their extinction, without denying that a combination of other factors has contributed to the erosion of their vital existential fibre'.[10]

In 1919, Gusinde found only 279 remaining Selk'nam on the Isla Grande. In a visit to Harberton that year, he was told there were only 100 Yagáns left. Gusinde made it his life's work to record the language, culture and beliefs of the peoples he believed were on the brink of disappearance. Equipped with a first-aid kit, which he used to cure wounds and minor ailments and gain the trust of his subjects, he visited Selk'nam and Yagán camps and later recalled how Tierra del Fuego's peoples came to regard him as 'the only white man that showed them charity'. Like Thomas Bridges before him, Gusinde learned the Yagán and the Selk'nam languages. He sat in at fireside chats and listened to gossip, myths, legends and snatches of history. Gusinde was invited to take part in the Selk'nam Hain ceremony—a complex combination of initiation ritual, theatre, mythological re-enactment, games and entertainment—and he also participated in the Yagán Kina initiation ceremony. He was also allowed to photograph these ceremonies, and these photographs were included in his four-volume magnum opus *Die Feuerland Indianer* (Fireland Indians, 1931–74)—the most comprehensive documentary record of Tierra del Fuego's Indigenous Peoples in existence. These books had a long and difficult gestation. The first volume was published in 1931, and many of Gusinde's notes for the subsequent volumes were destroyed in the war and written from memory, so that the final volume was not published until 1974, after his death.

Today, Gusinde's haunting pictures of Selk'nam and Yagán initiates with their painted bodies and surrealistic wooden masks have become part of the touristic iconography of southern Patagonia. His work is also honoured in the Martín Gusinde Anthropological Museum—the 'southernmost museum on earth', which was opened in Puerto Williams in 1975 and was subsequently remodelled into the L-shaped glass and wooden building at the edge of the town, with the carcass of a whale and a wooden Yagán 'wigwam' outside the main entrance. When I visited the museum, it was temporarily closed, but the director allowed me to look around. I spent most of the morning engrossed in the displays of Yagán tools and basketry, in Gusinde's marvellous photographs of the Kina ceremony and an enormous map showing the archipelago around Isla Navarino and the Beagle Channel, with its original Yagán names. The museum's director Alberto Serrano Fillol is in no doubt about Gusinde's contribution to the survival of Yagán culture. 'When Gusinde arrived here this world was very damaged,' he told me. 'He managed to enter it, and initiated a process of documentation that is still relevant for the communities here more than 100 years later. He's practically the only written source of this very profound environment.'

In societies that have been 'badly damaged by colonization, by the transformation of genocide, and the transformation of culture', Serrano argues that Gusinde's books, photographs and recordings—and the museum itself—can help recover that 'profound environment'. Today, descendants of Yagáns on the island regularly visit the museum to learn the songs and sacred ceremonies of their ancestors that had fallen into disuse. Gusinde was very much a practitioner of what is sometimes disparagingly called 'salvage' anthropology—the collecting and documenting of 'vanishing' Indigenous Peoples supposedly in the throes of a fatal contact with modernity. This practice has been critiqued by anthropologists and by Indigenous Peoples themselves as an intellectual extension of colonialism, which takes Indigenous cultural decline as a fait accompli and treats living peoples as if they were nothing more than disappearing relics of the past. But in a poignant historical irony, the missionary-anthropologist who set out to preserve what he believed was a dying Indigenous culture has left a resource that even young members of the Yagán community in Puerto Williams can now draw on as they seek to recover an identity that was once considered a source of shame.

* * *

Other visitors to Tierra del Fuego also saw its Indigenous Peoples as disappearing cultures to be 'salvaged', even if none of them did so as comprehensively as Gusinde. In Father Agostini's 1933 silent documentary, a Tehuelche family cheerfully speaks to camera, accompanied by the subtitle 'in the Patagonian plains the last Tehuelche Indians are dying'. In another sequence, one of Agostini's Selk'nam friends appears firing a bow and arrow as an example of 'one of the last of the Ona people'. In 1946, the Chilean experimental biologist Alejandro Lipschütz interviewed 'Indian hunters' in Tierra del Fuego and photographed its Indigenous Peoples for his study *Los últimos fueguinos* (The last Fuegians), commissioned by the Chilean government. The Franco-American anthropologist Anne Chapman (1922–2010) was a student of Claude Lévi-Strauss before she was introduced to Tierra del Fuego by the renowned palaeo-archaeologist Annette Laming-Emperaire in 1964. Chapman befriended the Selk'nam shaman Lola Kiepja (1874–1966), who was then ninety years old, and Chapman recalled how 'Lola greatly impressed me because she had not been vanquished by the tragedies of her life, because of her spontaneous laughter, her earnest, penetrating expression and because she had been the fountainhead of Selk'nam knowledge.'[11]

Chapman returned to Tierra del Fuego on various occasions and combined a professional interest in the First Nations of Tierra del Fuego with a strong emotional attachment to its people. Like Gusinde, Chapman set out to document a 'vanishing' culture and society through a stream of books, audio recordings, photographs and ethnological films. She also refuted the negative portrayals of the Yagán people, for which she held Darwin primarily responsible. In 2010, the year of her death, she published *European Encounters with the Yamana People of Cape Horn, before and after Darwin*, which she dedicated to the 'Jemmy Buttons of the World and their friends'. Chapman was withering in her condemnation of Darwin's 'ethnological meditations' and painstakingly picked apart his erroneous judgements of the Yagán language and culture. She berated Darwin for his lack of empathy and questioned whether he had actually witnessed some of the scenes he described. Chapman rejected the stereotypical representation of the Yagáns as 'the most defiled people in the world' and described them as 'fellow human beings, on a par with the well-known personages who encountered them'. 'Far from being wretched,' she argued, the Yagáns, 'lacked nothing in human terms. They experienced more than their share of problems owing to the exigencies of an often hostile environment and the threat of starvation.'[12]

When Chapman wrote these words, the material basis of Yagán culture and society had ceased to exist. All Yagáns wore clothes. They no longer hunted or used canoes. Most of them were Christianized, and many of them had intermarried with whites and had Spanish names. Most only spoke Spanish, and only a few still spoke the Yagán language or recalled the myths and ceremonies of their grandparents. The Kawésqar and Selk'nam were also subject to the same process of deculturation and assimilation, while the Haush—the 'devils' whom Darwin described in 1832—had disappeared completely. In 1974, the Selk'nam speaker Ángela Loij López, Chapman's friend and principal informant after the death of Lola Kiepja, died and earned the title of 'the last Selk'nam'. Chapman's Yagán friends on Navarino Island included the late Cristina Calderón (1928–2022), the last 'pure-blooded' Yagán who still spoke the Yagán language fluently. *Abuela* (Grandma) Cristina was a much-loved figure in the Yaghan community in the Villa Ukika neighbourhood in Puerto Williams as an ethnographer, craftswoman and educator. In her last years, she was regularly seen attending the religious services of all the different churches in Puerto Williams, sometimes going to various services on the same day. When she died in 2022 of complications from COVID-19 at the age of ninety-three, the governor of the Magallanes region declared a three-day mourning period, and newspapers across the world reported the passing of the 'last Yagán' or the 'last Fuegian'.[13]

The Tehuelche have often been the object of similar claims. In 1939, the Argentinian magazine *Caras y Caretas* described the death of the Aónikenk cacique Copacho (Kop'achus) as the 'last Tehuelche of southern Patagonia'.[14] In 1953, a Tehuelche peon named Rafael Capipe Huasque became 'the last Tehuelche' when he died in the corridor of a hospital in Río Gallegos after a lifetime of rural poverty and alcoholism.[15] And in 2019 the last known speaker of the Tehuelche language, Dora Manchado, died at the age of eighty-six in Río Gallegos; she was given the title of 'last Tehuelche'.[16] These iterations of the 'last Indian' share a very nineteenth-century tendency to seek the confirmation of cultural extinction in the death of a single individual. But human societies do not end like the dodo, the northern white rhinoceros or the Galapagos tortoise. The 2017 Chilean census listed 1,600 Yagáns in Chilean national territory, ninety-four of whom lived on Isla Navarino. The census also recorded 1,152 Selk'nam and 3,500 Kawésqar. In September 2023, the Chilean National Congress officially recognized the Selk'nam, the Yagán and the Kawésqar peoples as *pueblos australes* (southern peoples) and the original inhabitants of the lands they occupied.[17] Today, according to the 2022 Argentinian census, there are 17,420 Tehuelche and 145,783 Mapuche, some 60 per cent of whom live in Patagonia.[18]

* * *

How can the Indigenous Peoples the nineteenth century once believed were condemned to extinction still be part of the twenty-first? This resurgence is not as miraculous as it might seem. In the wake of the Conquest of the Desert, it was simply assumed that the Tehuelche and Mapuche would cease to exist or merge with the society of their conquerors because they were no longer visible as distinct separate societies. It was not until 1966–8 that Argentina began to count its Indigenous Peoples as a distinct category once again. In 2001, the National Census calculated its Indigenous Peoples according to whether one person per household claimed Indigenous heritage. In 2022, respondents were allowed to define themselves as Indigenous for the first time. In the months leading up to the census, Indigenous organizations campaigned around the slogan 'Yo soy de raíces indígenas' (I am of indigenous roots) in order to encourage respondents to embrace their heritage and state which group they belonged to.[19]

A similar process has unfolded in Chile, where official recognition and self-identification have given a new visibility to their 'extinct' First Peoples. In a 2001 paper on 'The Return of the Native', the late Mexican sociologist, anthropologist and advocate of Indigenous rights Rodolfo Stavenhagen observed that 'indigenous peoples have re-emerged in recent decades as new historical subjects, assertive actors in those fragile and incomplete democracies that brave the tempests of globalisation' after many years in which Latin America's ruling classes 'have always been quite happy to build nations without Indians'.[20] Today, both Argentina and Chile are attempting to build nations with Indians, and with their descendants, and this is one reason why the commemorations of the 'last' Tehuelche or Selk'nam were premature. The disappearance of a language is a tragic milestone for any society, which signifies the loss of a cultural world and all the feelings, social relationships and associations that are part of it. But in a world where 'identity' has become a fluid search for meaning that spans the past and present, human societies can also evolve and acquire culture while retaining or rediscovering aspects of the culture they have lost.

In both Argentina and Chile, the resurgence of Indigenous People as political and cultural protagonists is a story of survival and resilience, of a struggle for visibility, justice and reparation. In May 2023, I spoke to four members of the Consejo Consultivo y Participativo de los Pueblos Indígenas (Indigenous Peoples' Consultative and Participatory Council)—an umbrella organization containing representatives from across Argentina's Indigenous groups—in an improvised shelter made from tarpaulins and crates opposite Argentina's presidential palace in the Plaza de Mayo, Buenos Aires. This permanent camp is part of an ongoing protest calling for 'dialogue' between the government and the country's Indigenous Peoples, which had been in the square for more than a year. As we sat in the stifling Buenos Aires heat, a young Mapuche named Martín from northern Patagonia denounced the Argentinian parliament as a 'space of power' where 'the descendants of the genocide that killed our people continue to oppress us'.

Martín and his companions listed well-documented grievances experienced by many of Argentina's First Peoples, from land claims and evictions to the absence of birth certificates and ID cards, which effectively turns Indigenous Peoples into non-citizens and deprives them of access to state services. Estimates of Argentina's undocumented peoples have ranged from 300,000 to more than a million, most of whom are members of marginalized rural communities or inhabitants of urban *villas miserias* (shantytowns). Without an ID card, it is impossible to access a range of public services. In some Indigenous communities in the north of the country, undocumented children have died from illness and malnutrition, and there have also been reports of kidnapping and sexual abuse, which Martín compared to the forced separation of Indigenous children from their families during the Conquest of the Desert.

Evidence of such marginalization is not hard to find in Patagonia. Take a bus on the immaculately paved roads that connect Bariloche and the holiday villas around Lake Mascardi, and it's easy to miss the Ranquel and Mapuche settlements in the territories that Moreno donated to the nation as a national park. At the Arelauquen Country Club, on the shores of Lake Guttiérez, a wire fence encloses houses and sports facilities that cater to millionaire footballers, tennis players, bankers, lawyers and politicians. Until recently, the fence also enclosed the Lof Che José Celestino Quijada Mapuche community, which has been locked in dispute with Arelauquen's owners since 2007, when the club extended its wire fence, cutting the settlement off from the main road, its water supply and its cemetery. In December 2023, a federal judge ordered Arelauquen to remove the fence and return 35 hectares to the community following a prolonged lawsuit facilitated by the INAI.

Such disputes are rarely resolved in favour of the Mapuche. In principle, the 1994 amendment to the constitution of Argentina recognizes the general rights of Argentina's Indigenous communities to the 'possession and community property over lands they have traditionally occupied'. In practice, Indigenous claims based on customary or traditional land usage are not always recognized, particularly when these communities come up against the giant corporations and wealthy landowners who have replaced the cattle ranchers and sheep farmers as the new 'conquerors' of twenty-first-century Patagonia. In the red sandstone shale oil and gas deposits of Vaca Muerta in northern Patagonia, Mapuche communities have combined legal actions and protests in what has so far been a vain attempt to stop giant corporations that include the state-owned energy company YPF, Chevron and ExxonMobil from fracking operations that the Mapuche claim have drained and contaminated the water supply, caused tremors and driven away wildlife.

One of the longest-running Mapuche land disputes involves the Benetton clothing company. In 1991, Benetton bought an enormous ranch at Leleque, in the department of Cushamen, near Esquel, from the Compañia de Tierras Sud Argentino—the successor to the British-owned Argentine Southern Land Company, which was one of the main beneficiaries of Roca's Conquest of the Desert. In doing so, the company acquired nearly 900,000 hectares and overnight became the largest landowner in Argentina. In 2001, a group of Mapuche families occupied 385 hectares of unoccupied land on these holdings, which they 'recuperated' as ancestral territory. Thus began a dispute that has yet to be resolved. The Benetton family rejected these claims and defended its case in the courts. On 10 January 2017, more than 200 police raided one of the Mapuche 'Communities in Resistance' at the Leleque ranch with horses, helicopters and water cannon and evicted its occupants. In August that year, the National Gendarmes clashed with another Mapuche-Tehuelche community in Cushamen, in which an Argentinian activist Santiago Maldonado went missing and subsequently turned up dead, apparently as a result of drowning.

This was not good publicity for a company whose brand was based on its commitment to diversity capitalism, and in 2000 Benetton took the unusual decision to open a museum on the ranch dedicated entirely to the Tehuelche. The aim of this museum, according to its publicity brochure, was to create a 'meeting point in the heart of Patagonia, just as this land has been, for centuries, a place of encounter, conflict, trade and interaction between different populations'.[21] This 'meeting point' is located about an hour from Esquel along the Route 40, where a large sign bearing an image of the Tehuelche chief Copacho announces the 'Museo Leleque: Patagonia; Its History'. A tree-lined drive leads to the two reconstituted farm buildings, one of which houses some 14,000 Indigenous artefacts originally collected by a Ukrainian-Argentinian collector named Pablo Korchenewski. The organization of these exhibits follows the familiar sequential Patagonian museum trajectory, from prehistory to modernity, culminating in a section on 'Settlement, Sedentarism and Decadence', which explains the disappearance of the Tehuelche through poverty, alcoholism and illness.

This 'script' is illustrated with a meticulous reconstruction of a Tehuelche *toldo* and with Indigenous artefacts, maps, photographs, rifles and frontier paraphernalia. But the key to its intentions can be found in the reconstructed *boliche* (saloon bar-cum-café) in the adjoining building, where a framed interview with the late palaeontologist and archaeologist Rodolfo Casamiquela (1932–2008) proclaims 'Los mapuches verdaderos son muy pocos' (There are very few true Mapuche). A Tehuelche speaker and the author of twenty-four books, Casamiquela curated the museum's collections, and he was also known for his dogmatic insistence on the Chilean origins of Argentina's Mapuche peoples. His persistent dismissals of the Mapuche as 'foreign' intruders and invaders often turned his lectures and public appearances into objects of noisy protests or *escraches* from Mapuche and pro-Mapuche activists. In the interview, Casamiquela describes the Tehuelche as the authentic Indigenous inhabitants of Patagonia and the Mapuche as extraneous intruders from Chile, whose land claims lack legitimacy.

This is a common response in the rejection of Mapuche land claims in Patagonia, and the Leleque museum is part of that rejection. Just a few miles down the road from the museum, I stopped off at the Santa Rosa Leleque Mapuche community—a Mapuche squatter settlement on the Route 40. Beneath the Mapuche-Tehuelche flag, a sign described the site as 'Recuperated Ancestral Territory'. The settlement consists of a handful of log cabins, and it seemed to be deserted until I came across an elderly Mapuche man with a lined, leathery face and rheumy eyes who was walking his dogs. He talked cautiously about the complicated history of the community's dispute with Benetton, and I asked him whether he had documentary evidence to support his claims. 'Look at me,' he said. 'I'm a document. My grandfather and great-grandfather lived in this territory. I have Mapuche blood in my veins. What am I, if not a Mapuche?'

The Ranquel chief Mariano Rosas had once made much the same arguments in his discussions about land tenure with Mansilla in 1870, and as I walked away from the sign that read 'We resist, and we are here', it seemed to me that this message was the story of Patagonia's Indigenous Peoples. Depicted as primitive savages and relics of prehistory in the nineteenth century, their lands had been transformed from 'deserts' into investments and suppliers to global markets. Now twenty-first-century Patagonia was being re-imagined and re-configured as a source of sweaters, oil, gas, logs, salmon or hydropower, or as a playground for millionaires and recreational tourism, and its Indigenous Peoples remained marginalized outsiders in the lands of their ancestors. In 2007, 144 countries adopted the United Nations Declaration on the Rights of Indigenous Peoples, considered to be the most comprehensive legal guarantor of Indigenous rights in history, which aimed to establish 'minimum standards for the survival, dignity, wellbeing and rights of the world's indigenous peoples'.[22]

These 'minimum standards' have not always been met in Argentina. But today, Patagonia's First Peoples have allies that were not available to them in the nineteenth century. In Patagonia, as in many other countries, their history is being re-remembered and re-written; generals and politicians once hailed as heroes through the role they played in their destruction and defeat are being taken down from their pedestals. And in these circumstances, it remains to be seen whether Darwin's savages can find their way to a better future in the two states that conquered them on behalf of civilization and science and once believed they had disappeared.

EPILOGUE

THE HOUSE OF THE WIND

In April 2019, King Harald V and Queen Sonja of Norway visited Puerto Williams, on Navarino Island, during a state visit to Chile. To their surprise, the royal couple were greeted by the ninety-year-old Cristina Calderón and the president of the Bahía Mejillones Yagán community, David Alday, who presented them with a letter protesting the expansion of Norwegian-owned fish-farming operations into the Strait of Magellan and the Beagle Channel. The letter described salmon aquaculture as an 'environmental catastrophe' and urged the royal couple to halt its extension on behalf of 'the southernmost culture in the world' that was 'still resisting the consequences of the colonization process'.[1] This polite protest was part of a regional campaign against industrialized salmon farming in Argentina and Chile that included the Martín Gusinde Museum, the Yagán community of Bahía Mejillones, the local government, Greenpeace and the Omora Foundation—a non-profit scientific organization with a research station in Puerto Williams.

That same year, the government of Argentinian Tierra del Fuego called a temporary halt to the salmon farming project in the Beagle Channel, and in 2021 the Tierra del Fuego legislature approved a bill prohibiting all coastal salmon farming. The participation of the Yagán community in this campaign was remarkable for many reasons. As a result of the colonization of Tierra del Fuego by missionaries and its subsequent absorption into Chile and Argentina, the Yagáns effectively lost contact with the sea, and the Chilean and Argentinian naval presence further restricted their right to free navigation. And yet, in 2019, a community that so many observers believed was close to 'extinction' joined a conservation campaign that extended

271

from the Isla Navarino to Ushuaia and Punta Arenas, in which they reasserted their identity as the 'southernmost culture in the world' and declared themselves to be the traditional guardians of the maritime territories that their ancestors had once occupied.

This process of cultural rediscovery continued during the COVID-19 pandemic among the Yagán communities under quarantine in Puerto Williams and Bahía Mejillones.[2] In effect, the Indigenous Peoples of the 'end of the world' joined an international phenomenon of Indigenous environmental activism and cultural resurgence that has gained in strength and visibility in the twenty-first century. From Tierra del Fuego and the Amazon to Honduras to Standing Rock; from the tribal communities of Central India to the Mbororo pastoralist people in Chad and the Maasai Mara game reserve in Kenya—across the world, Indigenous Peoples have become political actors in defence of their cultural traditions and their homelands against the environmental depredations of a twenty-first-century capitalism that is no less destructive than its previous iterations.

Though the world's 476 million Indigenous Peoples make up 6.2 per cent of the global population, they account for 19 per cent of the extreme poor, according to the World Bank, and their life expectancy is nearly twenty years lower than the life expectancy of non-Indigenous Peoples.[3] Yet Indigenous Peoples also manage or exercise customary ownership over a quarter of the world's surface, containing some 80 per cent of global biodiversity. Their homelands have often been at the receiving end of extractivist industrial operations such as oil, gas, mining and logging, which have degraded and damaged the natural resources on which their culture and survival depends.

The Indigenous Peoples of Tierra del Fuego and Patagonia have been subject to these processes even before the colonization of their territories began. In the late nineteenth century, European and American sealers decimated the sea lions and fur seals that had been a staple of the Fuegian diet for thousands of years as well as depleting the whale population that had sometimes supplemented it. On the Isla Grande, the guanaco were culled to the brink of extinction and continue to be regularly poached despite the attempts by the local authorities to prevent it. In 1946, the Perónist military government introduced twenty Canadian beavers into Lake Fagnano on the Isla Grande in an attempt to create a local fur industry that would encourage settlement on the still sparsely populated Isla Grande. The introduction of the beavers into the island by the Argentinian navy was filmed by the state-controlled news agency and shown in cinemas across the country, where it was depicted as a transformational event that would bring prosperity to the whole country.[4] The fur industry never took off, but the beavers became an environmental catastrophe, which is still ongoing. There are now an estimated 100,000 beavers building dams in Tierra del Fuego and 60,000 in the Isla Navarino. Roads have been flooded, and whole swathes of forests have been turned into bogland, littered with the stumps and remains of trees used to construct dams and re-route streams and rivers.

The Patagonian mainland has also suffered environmentally from the consequences of large-scale livestock production. Some 30 per cent of Patagonian grasslands suffer from severe desertification as a result of overgrazing by sheep and cattle—exacerbated by wind erosion and wildfires—and more than 90 per cent of the territory is subject to soil degradation. Where the Patagonian 'desert' was once conquered on behalf of the sheep and cattle industry, large swathes of northern Patagonia are being reconfigured in accordance with twenty-first-century commodity extraction, through mining and hydrocarbon activities, logging and oil and shale gas production. In 2004, Jorge Nawel, the head of the Mapuche Confederation of Neuquén delivered a letter to the Securities Exchange Commission in New York, calling on the commission to investigate the 'consequences of uncontrolled exploitation' of hydrocarbons in the Vaca Muerta by companies listed on the exchange. 'Our culture is threatened, our territories are invaded and contaminated, our flora and fauna are poisoned, our air is affected by chemicals and our soil is shaking,' the letter informed the commission.[5]

In March 2021, wildfires devastated La Comarca Andina in the Argentine Patagonian Andes, burning 54,000 acres and destroying 100 houses. According to the Argentine Environment Agency, 93 per cent of these fires are due to human activity, much of which is the result of the reforestation programme carried out in the 1970s that brought faster-growing pine trees to the region. While these trees benefitted the logging industry, they also burned more easily. Without rebalancing these forests with less flammable species, *National Geographic* has predicted that 'Patagonia faces a fiery future filled with environmental degradation.'[6]

Few of the nineteenth-century scientists who predicted the imminent disappearance of the world's 'savages' and 'primitives' imagined that these same peoples might have knowledge and expertise that could be useful and even essential to humanity's survival. Dazzled by Europe's technological, military and political achievements, too many intellectuals saw such peoples as the collateral damage of civilization, whose evolutionary or cultural retardation condemned them to extinction in a racialized 'struggle for existence'. In applying Darwinian concepts to explain the dire effects of settler-colonialism, science, in effect, became an instrument in the construction of the Indigenous 'Other' and a mirror in which civilization could proclaim its cultural or racial superiority. These 'scientific' nostrums of civilizational progress and racial competition were also applied by the *criollo* governments of post-independence Latin America to the Indigenous 'barbarians' within their national borders.

Traces of that past can still be found in contemporary Latin America. In an article for *Harper's Magazine* in 1992, the novelist Mario Vargas Llosa described 'Indian peasants' in Peru as hermetically sealed communities who 'live in such a primitive way that communication is impossible'. Only through migration to the cities could they 'mingle with the other Peru' and renounce 'their culture, their language; their beliefs, their traditions and customs, and the adoption of the culture of their ancient masters' until 'after one generation they become mestizos. They are no longer Indian.'[7] In 2009, the Peruvian national newspaper *Correo* described Indigenous protestors blockading a road in the Peruvian Amazon as 'savages,'palaeolithic' and 'primitive' and called on the government to bomb them with napalm. Survival International cited this article as part of its 'Stamp It Out' campaign, which aims to counter: racist descriptions of tribal peoples in the media with the message 'proud not primitive'.[8] The campaign condemns language used to 'justify the persecution or forced "development" of tribal peoples'. It singles out '[t]erms like "stone age" and "primitive" [which] have been used to describe tribal people since the colonial era, reinforcing the idea that they have not changed and that they are backward'.[9]

It is nearly 200 years since Darwin described the First Peoples of Tierra del Fuego in much these terms. I thought of Darwin, FitzRoy and the *Beagle* when I drove from Puerto Williams to the Yagán cemetery at Bahía Mejillones along the single dirt road that traverses the island parallel to the coast. It was difficult to imagine the oppressive desolation that had destroyed Pringle Stokes as I passed the burnished sunlit surface of Onashaga—the Yagán name for the Beagle Channel. As I looked out at the deserted stony coves and beaches where the Yagáns built their shelter for thousands of years, Tierra del Fuego did not feel like the end of the world but a place of unique beauty and splendour that I was privileged to witness. It was a rare windless day, and the green and red tree-covered mountain slopes occasionally gave way to grey swathes of ghost forest devastated by beavers. The 'Indigenous Cemetery' was established by missionaries in the nineteenth century to aid the process of assimilation. Previously, the Yagán left their dead covered with tree bark or piled with stones on different islands, and overcoming these 'pagan' rituals was part of the preparation for a Christian afterlife—and also for the new life that civilization demanded of them.

In the 1930s, the cemetery was dug up by the Chilean navy and re-established in the same place, though the bodies and the names on the wooden crosses never correlated afterwards. Today, the cemetery is largely grown over, and piles of crosses are still stacked against the low picket fence that encloses it, and no one really knows which graves they belong to. At the upper part of the little burial ground, a shrine bearing the photograph of Cristina Calderón is planted with flowers, alongside an embroidered anchor beside her grave—a tribute by the Chilean navy. A smaller shrine marks the burial place of Calderón's son Martín, who, like his mother, died of COVID. This cemetery also belonged to the chain of historical events that connect the grave of the 'last Yagán' all the way back to Darwin, FitzRoy and Button.

Unlike the Tehuelche or the Mapuche, the Fuegian canoeists had no resources to be conquered or lands to be taken. Isolated from the wider world, they were able to survive in a landscape that most Europeans believed to be unfit for humans. Like every human society anywhere, they evolved a language and culture and a set of relationships with the land and with each other until, by a fluke of history, they became 'Darwin's savages'—objects of pity, contempt and scientific discussion in faraway conversations between learned men who debated whether they were prehistoric relics, the links between humans and apes or heathens in need of a salvation they had not asked for. In one of the shamanic songs recorded by Anne Chapman, the Selk'nam shaman Lola Kiepja described herself 'singing in the house of the wind' as she followed the footsteps of her ancestors across Ham'nia—the place of power in the western sky.[10]

Today, those songs are no longer sung, and the culture that produced them can never entirely be brought back. But as I drove away from the graveyard and looked out towards the channel that Martens had once painted so beautifully, I imagined the Selk'nam in Furlong's photograph, walking in a line along the beaches of Tierra del Fuego, and the Yagáns in their canoes waving to the emissaries of civilization, and I seemed to hear the shouts of 'Yammerschooner!' echoing across the water, which Darwin and his companions had once heard and misunderstood, just as they had misunderstood so much else. I pictured the Fuegian on the rock whom Darwin had imagined as a creature sunk in some primordial agony. I thought of Jemmy Button and Fuegia Basket, and the skulls and skeletons of 'lost races' that made their way to La Plata and so many other museums, and that so many museums are now seeking to return to the lands where they came from. And I thought then, as I have often thought since, that the conquest of Patagonia is not just part of Chilean and Argentinian history—the colonization of these stupendous lands and the destruction of its peoples is part of the history of the modern world that we all inhabit. Darwin's savages— and the conclusions he drew from them—are part of that history. And even today, in an era in which science is more humble, and supposedly more enlightened in its attitude to race and culture, we continue to find new ways of imagining the Other, new ways of constructing the savage, new ways of denying the humanity of those whom we still prefer to believe are not like us.

LIST OF ILLUSTRATIONS

1. Illustrations of 'Fuegians' from the voyage of the *Beagle*. *Credit: Wellcome Collection.*

2. *Expedition in the deserts of the South* by Calixto Tagliabúe. *Credit: Museo Saavedra via Wikimedia Commons.*

3. Charles Darwin and his eldest son William Erasmus Darwin, 1842. *Credit: Wikimedia Commons.*

4. Juan Manuel de Rosas, Argentine dictator, 1829. Portrait by Arthur Onslow. *Credit: Wikimedia Commons.*

5. 'Wáki killing a Puma', illustration from *At Home with the Patagonians* (1871) by George Chaworth Musters. *Credit:Archivo General de la Nación Argentina, AR-AGN-AGN01-AGAS-rg-83651.*

6. *La vuelta del malón* by Ángel della Valle, 1892. *Credit: Colección del Museo Nacional de Bellas Artes, Buenos Aires, Argentina, N°6297.*

7. Statue of General Julio Argentino Roca in Bariloche, Argentina, June 2020. *Credit: Bariloche33 via Wikimedia Commons, CC BY-SA 4.0.*

8. *Indian Auxiliaries on Choele Choel* by Antonio Pozzo, *Album de Vistas Expedición al Río Negro* (1879). *Credit: Museo Roca, Buenos Aires. MR-ME 4340.*

9. Exhibition room at the Museo de La Plata. *Credit: Archivo de resguardo del Colectivo GUIAS (Grupo Universitario de Investigación en Antropología Social).*

10. Wives of Inacayal and Foyel, with other Indigenous prisoners. Subsequently published as part of a series on 'aboriginal iconography' by M. A. Vignati, in the *Revista Museo de la Plata,* 1942. *Credit: Archivo General de la Nación Argentina, AR-AGN-*

NOTES

INTRODUCTION: AT THE WORLD'S END

1. Richard Darwin Keynes (ed.), *Charles Darwin's Beagle Diary* (Cambridge: Cambridge University Press, 2001), 122.
2. Ibid., 444.
3. Aimé Césaire, *Discourse on Colonialism* (New York: Monthly Review Press, 1996), 56.
4. Juan Antonio Garretón, *Partes detallados de la expedición al desierto de Juan Manuel de Rosas en 1833* (Buenos Aires: Eudeba, 1975), 11. Author's translation.
5. See, for example, Richard Weikart, *From Darwin to Hitler: Evolutionary Ethics, Eugenics and Racism in Germany* (Basingstoke: Palgrave Macmillan, 2004).
6. John Desmond Bernal, *Science in History Volume 1* (London: Cameron Associates, 1954), 483–4.
7. Adrian Desmond and James Moore, *Darwin's Sacred Cause: Race, Slavery and the Quest for Human Origins* (London: Allen Lane, 2009).
8. Gillian Beer, *Open Fields: Science in Cultural Encounter* (Oxford: Clarendon Press, 1996), 78.
9. See the discussion on 'The Moral State of Tahiti—and of Darwin', in Stephen J. Gould, *The Mismeasure of Man* (New York: W.W. Norton & Co.,1996), 413–23.
10. Héctor Palma, *Savage and Civilized: Darwin, Fitz Roy and the Fuegians* (Buenos Aires: Editorial Biblos, 2020), 159.
11. Paul Theroux, *The Old Patagonian Express by Train through the Americas* (New York: Washington Square Press, 1979), 407.
12. Ibid., 421.
13. 'Shunned by the Government, Villaruel Stands by Her Comments on Racism', *Buenos Aires Herald*, 24 July 2024.
14. David Viñas, *Indios, ejército y frontera* (México, D.F.: La Flor Azul, 2021), 17.

1.   RES NULLIUS

1. Benjamin Morrell, *A Narrative of Four Voyages* (New York: J. & J. Harper, 1832), 41–2.

2. Antonio Pigafetta, *The First Voyage around the World 1519–1522*, ed. Theodore Cachey Jr (Toronto: University of Toronto Press, 2007), 12.

3. María Rosa Lida de Malkiel, 'Para la toponomia argentina', *Hispanic Review*, vol. 20, no. 4 (Oct. 1952), 321–3.

4. James Weddell, *A Voyage towards the South Pole, Performed in the Years 1822–24* (London: Longman, Rees, Orme, Brown, and Green, 1827), 174.

5. Richard Hakluyt (ed.), *The Principal Navigations, Voyages, Traffiques and Discoveries of the English Nation*, vol. 16 (Edinburgh: E. & G. Goldsmid, 1890), 13.

6. Daniel Webb (ed.), *Selections from M. Pauw* (Bath: R. Cruttwell, 1795), 84.

7. John Hawkeworth, *An Account of the Voyages Undertaken by the Order of His Present Majesty: For Making Discoveries in the Southern Hemisphere*, vol. 1 (Perth: R. Morrison Junior, 1789), 153.

8. Philip Parker King, *Narrative of the Surveying Voyages of His Majesty's Ships Adventure and Beagle between the Years 1826 and 1836*, vol. 1 (London: Henry Colburn, 1839), 180.

9. Theroux, *Old Patagonian Express*, 442.

10. Roberto Bolaño, 'Patagonia el último lugar del mapa', *El Mundo*, 2 Nov. 2001, author's translation, https://www.elmundo.es/viajes/2001/VI02/VI02-pag02.html

11. These distinctions are not definitive or straightforward. Many Indigenous groups identified in the sixteenth century were no longer present in the early nineteenth. Some Indigenous names were Hispanicized approximations. Other denominations revolve around the Mapuche word 'che' (people or peoples), which identified people according to where they were located. So 'Mapuche' means 'people of the land / earth' from the Mapudungun word 'Mapu'. Tehuelche means 'brave people' or 'people of the south'. Puelche means 'people of the east' and 'Picunche' means 'people of the north'. In the nineteenth century, the Indigenous Peoples of central Chile were increasingly referred to as Mapuche, 'Aucas' or 'Araucanians'.

12. Basilio Villarino, *Diario del piloto de la Real Armada* (Buenos Aires: Imprenta del Estado, 1837), 86.

13. Once again, these different denominations of 'Fuegians' vary considerably. The Yagán or Yahgan people was the name given by missionaries to the inhabitants of the islands south of the Isla Grade and Cape Horn, who called themselves the Yamana or Yámana in their own

language, meaning 'people'. The Selk'nam were often referred to as the Ona people in Argentina but not in Chile.

2.   PATAGONIAN ENCOUNTERS

1. Louis-Antoine de Bougainville, *A Voyage round the World* (London: Nourse, 1772), 172.
2. Weddell, *Voyage towards the South Pole*, 6–7.
3. Hakluyt, *Principal Navigations*, 155–6.
4. Ibid., 11.
5. John Callander, *Terra Australis Cognita or Voyages to the Terra Australis*, vol. 2 (Edinburgh: A. Donaldson, 1768), 308.
6. Thomas Falkner, *A Description of Patagonia and the Adjoining Parts of South America* (Hereford: C. Pugh, 1774), 107.
7. Bougainville, *Voyage round the World*, 25.
8. Quoted in Alfred J. Tapson, 'Indian Warfare on the Pampa during the Colonial Period', *The Hispanic American Historical Review*, vol. 42, no. 1 (Feb. 1962), 1–28.
9. Ibid.
10. Falkner, *Description of Patagonia*, 97.
11. John Augustus Beaumont, *Travels in Buenos Ayres, and the Adjacent Provinces of the Rio de la Plata* (London: J. Ridgway, 1828), 57.
12. Emeric Essex Vidal, *Picturesque Illustrations of Buenos Ayres and Montevideo* (London: R. Ackermann, 1820), 55.
13. Carlos Martínez Sarasola, *Nuestros Paisanos los Indios* (Buenos Aires: Editorial Del Nuevo Extremo, 2011), 132.
14. Woodbine Parish, *Buenos Ayres and the Provinces of the Rio de la Plata* (London: J. Murray, 1839), 128–9.
15. Francis Bond Head, *Some Rough Notes Taken during Some Rapid Rides across the Pampas and among the Andes* (London: J. Murray, 1826), 118.
16. Beaumont, *Travels in Buenos Ayres*, 54–5.
17. Quoted in Sarasola, *Nuestros Paisanos*, 159.

3.   SAVAGES, PRIMITIVES AND CANNIBALS

1. Parish, *Buenos Ayres*, 338.
2. Paul Halsall, 'Mandeville on Prester John', Internet Medieval Source Book, Fordham University, https://sourcebooks.fordham.edu/source/mandeville.asp#:~:text=The%20Sourcebook%20is%20a%20collection%20of%20public%20domain%20and
3. Contemporary civilization staff of Columbia College, Columbia University, *Introduction to Contemporary Civilization in the West: A Sourcebook*,

3rd edn, vol. 1 (New York: Columbia University Press, 1960), 523–9, https://sites.miamioh.edu/empire/files/2022/12/1545-Sepulveda-On-the-Just-Causes-for-War-against-the-Indians.pdf

4. *Buffon's Natural History*, vol. 4 (London: J.S. Barr, 1797), 313.

5. Charles White, *An Account of the Regular Gradation in Man* (London: C. Dilly, 1759), 134–5.

6. Bronwen Douglas and Chris Ballard, *Foreign Bodies: Oceania and the Science of Race 1750–1940* (Canberra: Australia National University Press, 2008), 41.

7. Quoted in Thomas Henry Huxley, *Critiques and Addresses* (London: Macmillan, 1873), 141–2.

8. Quoted in Gustav Jahoda, *Images of Savages: Ancient Roots of Modern Prejudice in Western Culture* (London: Routledge, 1999), 21.

9. Ibid., 21–2.

10. Johann Baptist von Spix and Carl Friedrich Philipp von Martius, *Travels in Brazil, in the Years 1817–1820, Undertaken by Command of His Majesty the King of Bavaria* (London: Longman, Hurst, Rees, Orme, Brown, and Green, 1824), 241.

11. Jacques Julien Houton de Labillardiere, *Voyage in Search of La Pérouse* (London: Printed for John Stockdale, 1800), v–vi.

12. David Weber, *Bárbaros: Spaniards and Their Savages in the Age of Enlightenment* (New Haven: Yale University Press, 2005), 31.

13. Georg Forster, *A Voyage round the World in His Britannic Majesty's Sloop, Resolution, Commanded by Capt. James Cook, during the Years 1772, 3, 4, and 5*, vol. 2 (London: Printed for B. White, 1777), 459.

14. Weber, *Bárbaros*, 39.

15. Captain P. Parker King, *Narrative of the Surveying Voyages of His Majesty's Ships Adventure and Beagle between the Years 1826 and 1836*, vol. 1 (London: Henry Colburn, 1839), 24.

16. Robert FitzRoy, *Narrative of the Surveying Voyages of His Majesty's Ships Adventure and Beagle between the years 1826 and 1836*, vol. 2 (London: Henry Colburn, 1839), 121.

17. Ibid., 10.

18. FitzRoy, *Narrative of the Surveying Voyages*, appendix to vol. 2, 142–7.

19. Keynes, *Charles Darwin's Beagle Diary*, 7.

20. Charles Darwin, *Journal of Researches into the Natural History and Geology of the Countries Visited by H.M.S. Beagle* (London: J. Murray, 1845), 207–8.

21. Letter to Charles Thomas Whitley, 9 Sept. 1831, Darwin Correspondence Project, 'Letter no. 121', https://www.darwinproject.ac.uk/letter/?docId=letters/DCP-LETT-121.xml

4.   THE MAN ON THE ROCK

1. Darwin, *Journal of Researches*, 85.
2. Letter to Caroline Darwin, 24 Oct.–24 Nov. 1832, Darwin Correspondence Project, 'Letter no. 188', https://www.darwinproject.ac.uk/letter/?docId=letters/DCP-LETT-188.xml
3. Letter to J.S. Henslow, 11 Apr. 1833, Darwin Correspondence Project, 'Letter no. 204', https://www.darwinproject.ac.uk/letter/?docId=letters/DCP-LETT-204.xml
4. Quoted in Nick Hazlewood, *Savage: Survival, Revenge and the Theory of Evolution* (London: Sceptre, 2000), 131.
5. Letter to Caroline Darwin, 30 Mar.–12 Apr. 1833, Darwin Correspondence Project, 'Letter no. 203', https://www.darwinproject.ac.uk/letter/?docId=letters/DCP-LETT-203.xml
6. Darwin, *Journal of Researches*, 214.
7. Michael K. Organ, 'Conrad Martens: Journal of a Voyage from England to Australia aboard HMS Beagle and HMS Hyacinth 1833–35', State Library of NSW Press, Sydney, Arts and Humanities Commons, 1994, 25–6, https://ro.uow.edu.au/asdpapers/123
8. Sylvia Iparraguirre, *Tierra del Fuego* (Willimantic, CT: Curbstone Press, 2000), 61.
9. Weddell, *Voyage towards the South Pole*, 6.
10. Parker King, *Narrative of the Surveying Voyages*, 76.
11. John Macdouall, *Narrative of a Voyage to Patagonia and Terra del Fuégo* (London: Renshaw and Rush, 1833), 160–1.
12. John Byron, *Byron's Narrative of the Loss of the Wager* (London: H. Leggatt & Co., 1832), 126–7. For an assessment of the veracity of the 'murder' and its appropriation by Darwin and other writers, see Joseph L. Yannielli, 'A Yahgan for the Killing', *The British Journal for the History of Science*, vol. 46, no. 3 (Sept. 2013), 414–43.
13. Janet Browne, *Charles Darwin: Voyaging* (London: Pimlico, 2003), 249.
14. Letter to Catherine Darwin, 6 Apr. 1834, Darwin Correspondence Project, 'Letter no. 242', https://www.darwinproject.ac.uk/letter/?docId=letters/DCP-LETT-242.xml
15. Letter to Charles Whitley, 23 July 1834, Darwin Correspondence Project, 'Letter no. 250', https://www.darwinproject.ac.uk/letter/?docId=letters/DCP-LETT-250.xml

5.   EXTERMINATION

1. Charles Lyell, *Principles of Geology*, vol. 2 (London: Murray, 1832), 175.
2. Peter Burnett, 'State of the State Address', 6 Jan. 1851, https://governors.library.ca.gov/addresses/s_01-Burnett2.html#:~:text=That%20

a%20war%20of%20extermination,wisdom%20of%20man%20
to%20avert

3. Eduardo Galeano, *Open Veins of Latin America* (New York: Monthly Review Press, 1997), 185–6.
4. Rosas, Order of the Day, 11 May 1833 from Antonio del Valle, *Recordando el Pasado*, 55.
5. Quoted in Victor Kiernan, *European Empires from Conquest to Collapse, 1815–1960* (Leicester: Fontan, 1982), 160.
6. Ibid., 183.
7. Ibid., 190.
8. Ibid., 180.
9. Ibid., 181–2.
10. Ibid., 169.
11. Darwin, *Journal of Researches*, 101.
12. Letter to William Darwin Fox, 25 Oct. 1833, Darwin Correspondence Project, 'Letter no. 223', https://www.darwinproject.ac.uk/letter/?docId=letters/DCP-LETT-223.xml

6. THE EMPIRE AND THE FLAG

1. Henry Mayhew and John Binny, *The Criminal Prisons of London and Scenes of Prison Life* (London: Griffin, Bohn, 1862), 5.
2. Ibid., iii.
3. Thomas MacCauley, 'On the Motion that East-India Company's Charter Bill Be Read a Second Time, HC Deb 10 July 1833', *Hansard*, vol. 19, 479–550; full speech available at: https://api.parliament.uk/historic-hansard/people/mr-thomas-macaulay/1833
4. Letter to Caroline Darwin, 20 Sept. 1833, Darwin Correspondence Project, 'Letter no. 215', https://www.darwinproject.ac.uk/letter/?docId=letters/DCP-LETT-215.xml
5. Letter to Caroline Darwin, 9–12 Aug. 1834, Darwin Correspondence Project, 'Letter no.253', https://www.darwinproject.ac.uk/letter/?docId=letters/DCP-LETT-253.xml
6. Darwin, *Journal of Researches*, 140.
7. Letter to Caroline Darwin, 23 Oct. 1833, Darwin Correspondence Project, 'Letter no. 222', https://www.darwinproject.ac.uk/letter/?docId=letters/DCP-LETT-222.xml
8. Otto von Kotzebue, *A New Voyage round the World*, vol. 1 (London: H. Colburn & R. Bentley, 1830), 169.
9. See Robert FitzRoy and Charles R. Darwin, 'A Letter Containing Remarks on the Moral State of Tahiti, New Zealand, &c.', *South African Christian Recorder*, vol. 2, no. 4 (Sept. 1836), 221–38,

https://darwin-online.org.uk/content/frameset?pageseq=1&itemID=F1640&viewtype=text

10. Darwin, *Journal of Researches*, 209.

11. Quoted in Jahoda, *Images of Savages*, 18.

12. Russell Thornton, *American Indian Holocaust and Survival: A Population History since 1492* (Norman, OK, 1987), 32.

13. *Teara: The Encyclopedia of New Zealand*, https://teara.govt.nz/en/graph/36364/maori-and-european-population-numbers-1840-1881

14. 'Report of the Parliamentary Select Committee on Aboriginal Tribes, Select Committee on Aboriginal Tribes (British Settlements)', House of Commons (London, 1837).

15. Charles Hamilton Smith, *The Natural History of the Human Species* (Edinburgh: W.H. Lizars, 1852), 168.

16. Darwin, *Journal of Researches*, 435.

17. Charles R. Darwin et al., 'Queries Respecting the Human Race, to Be Addressed to Travellers and Others: Drawn Up by a Committee of the British Association for the Advancement of Science, Appointed in 1839', Report of the BAAS, Glasgow meeting, Aug. 1840, 447–58, https://darwin-online.org.uk/content/frameset?pageseq=11&itemID=F1975&viewtype=side

18. Thomas Robert Malthus, *An Essay on the Principle of Population* (London: J. Johnson, 1798), 41–8.

19. Nora Barlow (ed.), *The Autobiography of Charles Darwin 1809–1882* (London: Collins, 1958), 88.

20. Paul H. Barrett et al. (eds), *Charles Darwin's Notebooks, 1836–1844* (Ithaca, NY: Cornell University Press, 1989), 465.

21. Ibid., 409.

7. THE CALIGULA OF THE RIVER PLATE

1. Quoted in William Dusenberry, 'Juan Manuel de Rosas as Viewed by Contemporary American Diplomats', *The Hispanic American Historical Review*, vol. 41, no. 4 (Nov. 1961), 495–514.

2. Waldo Ansaldi, *Rosas y su tiempo* (Buenos Aires: Centro Editor de América Latina, 1984), 1–2.

3. A British Gentleman, *Rosas, and Some of the Atrocities of His Dictatorship in the River Plate: In a Letter to the Right Honourable the Earl of Aberdeen, &c.* (London: Simmonds and Clowes, 1844). See also Ricardo D. Salvatore, 'De la ficción a la historia: El fusilamiento de indios de 1836', *Quinto Sol*, vol. 18, no. 2 (2014), 1–31.

4. Will MacCann, *Two Thousand Miles' Ride through the Argentine Provinces*, vol. 2 (London, 1853), 6.

5. For an English translation, see Angel Flores, 'El Matadero, Esteban Echeverria', *New Mexico Quarterly*, vol. 12, no. 4 (1942), 389–405, https://digitalrepository.unm.edu/cgi/viewcontent.cgi?article=2096&context=nmq

6. Domingo Sarmiento, *Facundo: Civilization and Barbarism* (Berkeley: University of California Press, 2004), 77.

7. Ibid., 45–6.

8. Quoted in Lucas Potenze, *Científicos y religiosos en Tierra del Fuego* (n.p.: Editoria Cultural Tierra del Fuego, 2021), 71. Author's translation.

9. Quoted in Nicolas Shumway, *The Invention of Argentina* (Berkeley: University of California Press, 1991), 141.

10. Colonel Anthony King, 'Rosas and the Argentine Republic', *The North American Review*, vol. 69, no. 144 (1849), 43.

11. Letter from B.J. Sulivan to Charles Darwin, 4 July 1845, Darwin Correspondence Project, 'Letter no. 886', https://www.darwinproject.ac.uk/letter/?docId=letters/DCP-LETT-886.xml

12. Darwin, *Journal of Researches*, 73.

13. Vicente Pérez Rosales, *Times Gone By: Memoirs of a Man of Action* (Oxford: Oxford University Press, 2003), 370.

8.   THE CITY AND THE DESERT

1. Benjamin Franklin Bourne, *The Captive in Patagonia: Or, Life among the Giants, a Personal Narrative* (Boston: Gould and Lincoln, 1853), 230.

2. Alberto Harambour, *Soberanías fronterizas: Estado y capital en la colonización de Patagonia (Argentina y Chile, 1880–1922)* (Valdivia: Ediciones Universidad Austral de Chile, 2019).

3. Quoted in John Marsh, *The Story of Commander Allen Gardiner* (London: J. Nisbet, 1877), 78.

4. Quoted in Hazlewood, *Savage*, 160.

5. William Parker Snow, 'A Few Remarks on the Wild Tribes of Tierra del Fuego from Personal Observations', *Transactions of the Ethnological Society of London*, vol. 1 (1861), 261–7.

6. O'Higgins/Smith correspondence available at Duncan Campbell and Gladys Grace website: https://www.patlibros.org/index.php?thm=init&fun=init&lan=eng&new=N&pol=Y#mk2div2. Author's translation.

7. For Brown's account of this episode, see Charles H. Brown, *Insurrection at Magellan: Narrative of the Imprisonment and Escape of Capt. Chas. H. Brown from the Chilean Convicts* (Boston: E.H. Appleton, 1854). For analysis of the wider geopolitical context of the mutiny, see: Manual Llorca-Jaña, 'Britain's Involvement in Chile's Cambiaso Mutiny, 1851–2: A Case of

Political Dependency at the Dawn of the Republic', *Itinerario*, vol. 47, no. 1 (2023), 40–58, https://www.cambridge.org/core/journals/itinerario/article/britains-involvement-in-chiles-cambiaso-mutiny-18512-a-case-of-political-dependency-at-the-dawn-of-the-republic/3 B044B488B102E871CF1DAC0C0D06FC8

9.   HINTERLANDS

1. See 'Constitución de la Nación Argentina de 1853', https://www.infoleg.gob.ar/?page_id=3873
2. From Pedro Cayuqueo, *Historia Secreta Mapuche*, vol. 1 (Santiago: Catalonia, 2019), 74.
3. Full letter in Antonio del Valle, *Recordando el pasado: 'Campañas por la civilización'*, vol. 1 (Buenos Aires: n.p., 1926), 22–7.
4. Auguste Guinnard, *Three Years' Slavery among the Patagonians* (London: R. Bentley and Son, 1871), 221–2.
5. For Spanish speakers, Carlos Sarasola Martínez's collection of letters written by Calfucurá and other Pampean caciques is an essential resource for students of nineteenth-century 'frontier diplomacy' in Argentina. See *La Argentina de los Caciques: O el país que no fue* (Buenos Aires: Editorial del Nuevo Extremo, 2014).
6. Frederick Jackson Turner, *The Frontier in American History* (New York: H. Holt and Co., 1920), 3.
7. Alberto Sarramone, *Historia del Antiguo Pago del Azul* (Azul: Editorial Biblos Azul, 1997), 29.
8. Del Valle, *Recordando el Pasado*, 12.
9. Francisco Moreno, *Reminiscencias de Perito Moreno* (Buenos Aires: Elefante Blanco, 1997), 100.
10. W.H. Hudson, *El Ombú* (London: Duckworth & Co., 1909), 57–8.
11. See Canal Encuentor, 'Antirracismo en tiempo presente: El color del antirracismo', YouTube, May 2022, https://youtu.be/FhFYOr-4rT8
12. For example, in Esteban Echeverría's epic narrative poem 'The Captive', where the protagonists of a captive-taking raid are depicted as blood-drinking vampires, *La cautiva* (Buenos Aires: Vaccaro, 1916).
13. For an English-language discussion of captive-taking in the Argentinian hinterland and its wider cultural associations, see Susana Rotker, *Captive Women: Oblivion and Memory in Argentina* (Minneapolis: University of Minnesota Press, 2002).
14. Jorge Luis Borges, 'Story of the Warrior and the Captive', from *Labyrinths: Selected Stories & Other Writings* (New York: New Directions, 1962).

15. R.B. Cunninghame-Graham, *Thirteen Stories* (London: William Heinemann, 1900), 165–6.

10. TRAVELLERS AND SETTLERS

1. Don Guillermo Cox, 'A Journey across the Southern Andes of Chile, with the Object of Opening a New Route across the Continent', *Proceedings of the Royal Geographical Society of London*, vol. 8, no. 5 (1863–4), 160–2.

2. George Chaworth Musters, *At Home with the Patagonians: A Year's Wanderings over Untrodden Ground from the Straits of Magellan to the Río Negro* (London: J. Murray, 1871), 185.

3. Julius Beerbohm, *Wanderings in Patagonia* (London: Chatto & Windus, 1879), 84.

4. Annie Allnutt Brassey, *A Voyage in the 'Sunbeam': Our Home on the Ocean for Eleven Months* (Chicago: Belford, Clarke & Co., 1881), 127.

5. Florence Dixie, *Across Patagonia* (London: R. Bentley, 1879), 84.

6. See 'Jean Raspail, French Writer and Hero of the Right Who "Invaded" the UK', *The Telegraph*, 28 June 2020.

7. Quoted in E.H. Bowen, 'The Welsh Colony in Patagonia 1865–1885: A Study in Historical Geography', *The Geographical Journal*, vol. 132, no. 1 (Mar. 1966), 16–27.

8. Glyn Williams, 'Welsh Settlers and Native Americans in Patagonia', *Journal of Latin American Studies*, vol. 11, no. 1 (May 1879), 41–66.

11. OF APES AND MEN

1. Letter to Charles Darwin, 31 Jan. 1862, Darwin Correspondence Project, 'Letter no. 3426', https://www.darwinproject.ac.uk/letter/?docId=letters/DCP-LETT-3426.xml

2. Letter to Charles Kingsley, 6 Feb. 1862, Darwin Correspondence Project, 'Letter no. 349', https://www.darwinproject.ac.uk/letter/?docId=letters/DCP-LETT-3439.xml

3. Charles Darwin, *The Origin of Species* (Hertfordshire: Wordsworth Editions, 1998), 30.

4. Robert Knox, *The Races of Man: A Fragment* (London: H. Renshaw, 1850), 30.

5. George Fitzhugh, *Sociology for the South: Or the Failure of Free Society* (Richmond, VA: A. Morris, 1854), 227.

6. Josiah Clark Nott et al., *Types of Mankind: Or, Ethnological Researches* (Philadelphia: J.B. Lippincott & Co., 1854), 79.

7. Quoted in John S. Haller, *Outcasts of Evolution: Scientific Attitudes of Racial Inferiority 1859–1900* (Urbana: University of Illinois Press, 1971), 140.
8. Ibid., 100.
9. For in-depth studies of nineteenth-century Victorian anthropology and its treatment of Indigenous Peoples, see George Stocking, *Victorian Anthropology* (New York: Free Press, 1991), and 'The Dark-Skinned Savage: The Image of Primitive Man in Evolutionary Anthropology', in George Stocking Jr, *Race, Culture, and Evolution: Essays in the History of Anthropology* (Chicago: University of Chicago Press, 1982).
10. John Lubbock, *The Origin of Civilisation and the Primitive Condition of Man* (London: Longmans, Green, 1870), v.
11. John Lubbock, *Pre-historic Times, as Illustrated by Ancient Remains, and the Manners and Customs of Modern Savages* (London: Williams and Norgate, 1865), 484–94.
12. James Hunt, 'The Extinction of Races, by Richard Lee', *Journal of the Anthropological Society of London*, vol. 2 (1864), xcv–xcix.
13. Gilbert Malcolm Sproat, *Scenes and Studies of Savage Life* (London: Smith, Elder and Co., 1868), 278.
14. Robert Brown, *The Races of Mankind: Being a Popular Description of the Characters, Manners, and Customs of the Principal Varieties of the Human Family*, vol. 3 (London: Cassell, Petter & Galpin, 1873), 199.
15. Lubbock, *Pre-historic Times*, 491.
16. See Thomas Henry Huxley, 'The Struggle for Existence', in *Evolution and Ethics, and Other Essays* (London: Macmillan, 1894), 202–18.
17. Alfred R. Wallace, 'The Origin of Human Races and the Antiquity of Man Deduced from the Theory of "Natural Selection"', *Journal of the Anthropological Society of London*, vol. 2 (1864), clviii–clxx.
18. Quoted in Haller, *Outcasts of Evolution*, 127.
19. Paul Topinard, *Anthropology* (London: Chapman and Hall, 1878), 528.
20. Letter to Thomas Bridges, 6 Jan. 1860, Darwin Correspondence Project, 'Letter no. 2640', https://www.darwinproject.ac.uk/letter/?docId=letters/DCP-LETT-2640.xml
21. J.P.M. Weale to Charles Darwin, 10 Dec. 1867, Darwin Correspondence Project, 'Letter no. 5722', https://www.darwinproject.ac.uk/letter/?docId=letters/DCP-LETT-5722.xml
22. W.E. Darwin to Charles Darwin, 27 June [July 1863?], Darwin Correspondence Project, 'Letter no. 4222F', https://www.darwinproject.ac.uk/letter/?docId=letters/DCP-LETT-4222F.xml
23. Quoted from 'Retratos literarios de Juan Manauel de Rosas', Historia Argentina, *La Gazeta Federal*, https://lagazeta.com.ar/retratos_rosas.htm

24. Charles Darwin, *Descent of Man, and Selection in Relation to Sex*, vol. 1 (London: J. Murray, 1871), 35.

25. Ibid., 178.

26. *The Descent of Man, and Selection in Relation to Sex*, vol. 2 (London: John Murray, 1871), 404–5.

27. The first six volumes of Brehm's popular zoological encyclopaedia were published in German in the 1860s. It is not clear how Darwin came across Brehm's story of the altruistic baboon, but he was familiar with his work. In 1868, he wrote to Brehm's German publisher for permission to use some of his illustrations in *Descent*. See Charles Darwin to Bibliographisches Institut, Darwin Correspondence Project, 'Letter no. 6235', https://www.darwinproject.ac.uk/letter/?docId=letters/DCP-LETT-6235.xml

12. THE BONE COLLECTORS

1. Cited in 'Buenos Ayres and the Republic of the Banda Oriental', *The American Whig Review*, vol. 3 (1846), 160–8.

2. For an account of these debates, see Alex Levine and Adriana Novoa, *¡Darwinistas!: The Construction of Evolutionary Thought in Nineteenth Century Argentina* (Leiden: Brill, 2012); also, Novoa and Levine, *From Man to Ape: Darwinism in Argentina, 1870–1920* (Chicago: University of Chicago Press, 2010).

3. W.H. Hudson, *Far Away and Long Ago: A History of My Early Life* (New York: E.P. Dutton and Company, 1918).

4. Ruth Tomalin, *W.H. Hudson: A Biography* (Oxford: Oxford University Press, 1984), 103.

5. W.H. Hudson, *Idle Days in Patagonia* (London: Dent, 1924), 217.

6. Ibid., 39.

7. J.G. Frazer, *The Scope of Social Anthropology: A Lecture Delivered before the University of Liverpool, May 14, 1908* (London: Macmillan and Co., 1908).

8. For full translation of Cuvier's anatomical report, see Marguerite Johnson and Alistair Rolls, 'Georges Cuvier's Autopsy Report on Sara Baartman: A Translation and Commentary', *The Journal of the Society for the History of Discoveries*, vol. 55, no. 2 (2023), 170–5.

9. Quoted in Desmond and Moore, *Darwin's Sacred Cause*, 34.

10. Quoted in José Luis Alonso Marchante, *Menéndez, rey de la Patagonia* (Santiago: Editorial Catalonia, 2014), 50.

11. Francisco P. Moreno, *Viaje a la Patagonia Austral* (Buenos Aires: Imprenta la Nación, 1879), 7.

12. Quoted in Levine and Novoa, *¡Darwinistas!*, 54.

13. Novoa and Levin, *Man to Ape*, 4.

14. Levine and Novoa, *¡Darwinistas!*, 54.
15. Ibid., 117.
16. Novoa and Levine, *Man to Ape*, 150.
17. Jens Andermann, 'Argentine Literature and the "Conquest of the Desert", 1872–1896', from 'Relics and Selves: Iconographies of the National in Argentina, Brazil and Chile, 1880–1890', digital exhibition: http://www7.bbk.ac.uk/cilavs/relics-and-selves-iconographies-of-the-national-in-argentina-brazil-and-chile-1880-1890/
18. Quoted in Edward Brudney, 'Manifest Destiny, the Frontier, and "El Indio" in Argentina's Conquista del Desierto', *Journal of Global South Studies*, vol. 36, no. 1 (Spring 2019), 116–44.

13. THE GREAT WALL OF ARGENTINA

1. Lucio V. Mansilla, *A Visit to the Ranquel Indians* (Lincoln, NE: University of Nebraska Press, 1997), 300.
2. Quoted in Walter Delrio and Pilar Pérez, 'Beyond the "Desert": Indigenous Genocide as a Structuring Event in Northern Patagonia', in Carolyne R. Larson (ed.), *The Conquest of the Desert: Argentina's Indigenous Peoples and the Battle for History* (Albuquerque: University of New Mexico Press, 2020), 127.
3. Quoted in Gabriel Cid, 'De la Araucanía a Lima: Los usos del concepto "civilización" el la expansión territorial del Estado chileno, 1855–1883', *Estudios Ibero-Americanos*, vol. 38, no. 2 (July–Dec. 2012), 265–83.
4. Ibid.
5. Sarasola, *Argentina de los caciques*, 131.
6. Álvaro Barros, *Fronteras y territorios federales de la Pampas del Sud* (Buenos Aires: Impr. de tipos á vapor, Belgrano, 1872), 190.
7. Exchange of letters in Manuel J. Olascoaga, *Estudio topográfico de la Pampa y Río Negro* (Buenos Aires: n.p., 1880), xv–xxx. Author's translation.
8. Quoted in Marcelo Valko, *Pedagogía de la desmemoria: Crónicas y estrategias del genocidio invisible* (Buenos Aires: Ediciones Continente, 2020),157.
9. Alfredo Ebelot, *Recuerdos y relatios de la Guerra de Fronteras* (Buenos Aires: Plus Ultra, 1968), 1.
10. Sarasola, *Argentina de los caciques*, 153.
11. Bernardo Arriaga, letter to Governor Aristóbulo del Valle, 9 May 1875, exhibit at Museo José A. Mulazzi, Tres Arroyos.
12. 'He visto á Pincén', *El Mosquito*, 15 Dec. 1878.
13. Ebelot, *Recuerdos y relatios*, 23.
14. Beatriz Celina Doallo, *Juan Manuel de Rosas: El exilio del restaurador 1852–1877* (Buenos Aires: Ediciones Fabro, 2012), 276.

PART III: CONQUEST

1. Quoted in Álvaro Yunque, *Calfucurá: La conquista de las pampas* (Buenos Aires: Biblioteca Nacional, 2008), 461. Author's translation.

14. THE CONQUEST OF THE DESERT

1. In July 2023, the Bariloche town council announced that it would take the Roca monument down, with the approval of the National Monuments Commission. In 2024, it was still there. On 24 Mar. that year—the date of Argentina's National Day of Remembrance for Truth and Justice, which commemorates the 1976 military coup—local Mapuche activists covered the statue with a replica of a giant *cultrún* (ceremonial drum).
2. Valko, *Pedagogía de la desmemoria*.
3. 'La verdad sobre la mentira mapuche', 22 Oct. 2021, Infobae, https://www.infobae.com/opinion/2021/10/22/la-verdad-sobre-la-mentira-mapuche/
4. Octavio Amadeo, *Vidas Argentinas* (Buenos Aires: Bernabé, 1934), 19.
5. For Roca's full address to Congress, see Olascoaga, *Estudio topográfico*, xxxiii–xxxvii.
6. Olascoaga, Estudio topográfico, 137.
7. Quoted in Gabriela Nouzeilles and Graciela Montaldo (eds), *The Argentina Reader* (Durham, NC: Duke University Press, 2002), 164.
8. Statistics from Julio Vezub and Mark Healey, '"Occupy Every Road and Prepare for Combat": Mapuche and Tehuelche Leaders Face the War in Patagonia', in Larson, *Conquest of the Desert*, 51–2.
9. Conrado Villegas, *Expedición al Gran Lago Nahuel Huapi en el Año 1881* (Buenos Aires: Editorial Universitaria de Buenos Aires, 1974), 23.
10. Letter in Williams, 'Welsh Settlers and Native Americans'.
11. Quoted in Pilar Pérez, *Archivos del silencio: Estado, indígena y violencia en Patagonia central 1878–1941* (Buenos Aires: Prometeo Libros, 2016), 93.
12. Olascoaga, *Estudio topográfico*, xvii.
13. James Scobie, *Argentina: A City and a Nation* (Oxford: Oxford University Press, 1871), 122.
14. Quoted in Pérez, *Archivos del silencio*, 88.
15. Walter Delrio et al., *En el País de Nomeacuerdo: Archivos y memorias del genocidio del Estado Argentino sobre los pueblos originarios, 1870–1950* (Viedma: Open Edition Books, 2018), 104.
16. Olascoaga, *Estudio topográfico*, xviii.
17. Both quotes from Walter Delrio et al., 'Discussing Indigenous Genocide in Argentina: Past, Present, and Consequences of Argentinian State

Policies toward Native Peoples', *Genocide Studies and Prevention*, vol. 5, no. 2 (Aug. 2010).

18. Marisa Malvestitti, *El parlamento imaginario de Ignacio Cañiumir* (Viedma, Río Negro: Fundación Ameghino, 1994).

19. Full Katrülaf testimony in Margarita Canio Llanquinao and Gabriel Pozo Menares, *Historia y conocimiento oral mapuche* (Santiago de Chile: Consejo Nacional de la Cultura y las Artes, 2013), 275–321.

20. Quoted in 'Mañana, primer centenario del adios a Sayhueque', *Río Negro*, 7 Sept. 2003, https://www.rionegro.com.ar/manana-primer-centenario-del-adios-a-sayhueque-HRHRN03090720072011/

15. THE TEMPLE OF EVOLUTION

1. Nora Barlow (ed.), *The Autobiography of Charles Darwin 1809–1882* (London: Collins, 1958), 80.

2. Quoted in Levine and Novoa, *¡Darwinistas!*, xiv–xv.

3. *Informe oficial de la Comisión científica agregada al Estado Mayor de la expedición as Río Negro (Patagonia) realizada en los meses de Abril, Mayo y Junio de 1879 bajo los ordenes de General D. Julio A. Roca* (Buenos Aires: Imprenta de Ostwald y Martinez, 1881).

4. To William Graham, 3 July 1881, Darwin Correspondence Project, 'Letter no. 13230', https://www.darwinproject.ac.uk/letter/?docId=letters/DCP-LETT-13230.xml

5. Ebelot, *Recuerdos y relatios*, 17.

6. Eduardo Ladislao Holmberg, *Carlos Roberto Darwin* (Buenos Aires: El Nacional, 1882), 192.

7. Quoted in Larson, *Conquest of the Desert*, 117.

8. Quoted in Novoa and Levine, *Man to Ape*, 84.

9. Clemente Onelli, *Trepando los Andes* (Buenos Aires: Continente, 1904), 227.

10. Quoted in Gustavo Vallejo, 'Museo y derechos humanos: Un templo de la ciencia finisecular en La Plat y aspectos de su relación con los pueblos orginarios', *Revista derecho y ciencias sociales*, vol. 7 (Oct. 2012), 146–64, https://revistas.unlp.edu.ar/dcs/article/view/11176/10220

11. Álvaro Bravo, *El Museo Vacío* (Buenos Aires: Eudeba, 2017). For further exploration of the relationship between museums and state expansion in Argentina, see Irina Podgorny and María Margaret Lopes, *El desierto en una vitrina: Museos e historia natural en la Argentina, 1810–1890* (México, D.F.: LIMUSA, 2008).

12. Florentino Ameghino, *Contribución al conocimiento de los mamíferos fosiles de la República Argentina* (Buenos Aires: Impr. de P.E. Coni é hijos, 1889), xv.

16. 'PRISONERS OF SCIENCE'

1. Full letter in Moreno, *Reminiscencias de Perito Moreno*, 177–81.
2. Author interview, 8 Mar. 2023. See also Máximo Ezequiel Farro, 'Historia de las colecciones en el Museo de La Plata 1884–1906: Naturalistas viajeros, coleccionistas y comerciantes de objetos de historia natural a fines del Siglo XIX' (PhD diss., Universidad Nacional de La Plata, Facultad de Ciencias Nacionales y Museo, 1 Jan. 2008), https://core.ac.uk/reader/19331504
3. Quoted in Colectivo GUIAS, *Antropología del genocidio: Identificación y restitución* (La Plata: De la Campana, 2010), 23.
4. Inacayal quote from Ricardo D. Salvatore, 'Live Indians in the Museum: Connecting Evolutionary Anthropology with the Conquest of the Desert', in Larson, *Conquest of the Desert*, 97–121.
5. Luis María Drago, *Los hombres de presa* (Buenos Aires: n.p., 1888), 141.
6. Carolina Vanegas Carrasco, 'Modelos involuntarios: Victor de Pol y los desafíos de la escultura de tema étnico en Argentina, Victor de Pol (1865–1925)', *MODOS: Revista de História da Arte*, vol. 5, no. 2 (2021), 16–35, https://typeset.io/papers/modelos-involuntarios-victor-de-pol-y-los-desafios-de-la-3394stw7ng
7. Karina Oldani, Miguel Añon Ssuarez and Fernando Miguel Pepe, 'Las muertes invisibilizadas del Museo de de La Plata', *Corpus: Archivos virtuales de la alteridad americana*, vol. 1, no. 1 (Jan.–June 2011).
8. Quoted in Alfredo Fernández, 'Inakayal y su agonía de 100 años', *Cosecha Roja*, 26 Aug. 2016. Author's translation, https://www.cosecharoja.org/inakayal-y-su-agonia-de-100-anos/
9. Quoted in Larson, *Conquest of the Desert*, 97.
10. 'Campsite Places Humans in Argentina 14,000 Years Ago', *Smithsonian Magazine*, 3 Oct. 2016, https://www.smithsonianmag.com/smart-news/campsite-suggests-humans-argentina-14000-years-ago-180960667/

17. FIN DEL MUNDO

1. Agustín de Vedia, *El teniente general Julio A. Roca y el comercio inglés: El gran banquete* (Buenos Aires: Imprenta de 'La Tribuna nacional', 1887).
2. Ingrid E. Fey, 'Peddling the Pampas: Argentina at the Paris Universal Exposition of 1889', in William H. Beezley and Linda A. Curcio-Nagy (eds), *Latin American Popular Culture since Independence* (Lanham, MD: Rowman & Littlefield, 2012), 100.
3. Ibid., 100–1.
4. William Eleroy Curtis, 'The Other End of the Hemisphere', *Harper's Magazine*, Nov. 1887.

5. Jean Baudrillard, 'Tierra del Fuego—New York', in *Screened Out*, trans. Chris Turner (London: Verso, 2002), 128–33.

6. Cited in Elizabeth Lydia Marsh Gardiner (ed.), *Records of the South American Missionary Society* (London: South American Mission Society, 1896), 41.

7. Lucas Bridges, *Uttermost Part of the Earth: The First History of Tierra del Fuego and the Fuegian Natives* (New York: Rookery Press, 1949), 39–40.

8. Ferdinand Hestermann and Martin Gusinde (eds), *Yamana–English: A Dictionary of the Speech of Tierra del Fuego* (Mödling: Missionsdruckerei St Gabriel, 1933).

9. See appendix to Thomas Bridges, *A Short Account of Tierra del Fuego and Its Inhabitants by Thomas Bridges* [writings compiled by his children, 1930], Duncan Campbell and Gladys Grace website, https://patlibros.org/tdf/doc.php#:~:text=The%20natives%20of%20Tierra%20del,is%20decidedly%20the%20most%20euphonious

10. Bridges, *Short Account of Tierra del Fuego*, 37.

11. Marchante, *Menéndez, rey de la Patagonia*, 44.

12. John Randolph Spears, *The Gold Diggings of Cape Horn: A Study of Life in Tierra del Fuego and Patagonia* (New York: G.P. Putnam's Sons, 1895), 23.

13. Ibid., 45.

14. Bridges, *Short Account of Tierra del Fuego*, 126.

15. Gardiner, *Records of the South American Missionary Society*, 30.

16. Cited in Robert Young, *From Cape Horn to Panama: A Narrative of Missionary Enterprise among the Neglected Races of South America, by the South American Missionary Society* (London: South American Mission Society, 1905), 68.

17. Spears, *Gold Diggings of Cape Horn*, 58.

18. Rockwell Kent, *Voyaging Southward from the Strait of Magellan* (New York: G.P. Putnam's Sons, 1924), 34.

19. Letter to Charles Darwin, 25 Dec. 1866, Darwin Correspondence Project, 'Letter no. 5325', https://www.darwinproject.ac.uk/letter/?docId=letters/DCP-LETT-5325.xml

20. Letter to Charles Darwin, 27 June 1870, Darwin Correspondence Project, 'Letter no. 7246', https://www.darwinproject.ac.uk/letter/?docId=letters/DCP-LETT-7246.xml

21. Letter to B.J. Sulivan, 30 June 1870, Darwin Correspondence Project, 'Letter no. 7256', https://www.darwinproject.ac.uk/letter/?docId=letters/DCP-LETT-7256.xml

22. Letter to Charles Darwin, 18 Mar. 1881, Darwin Correspondence Project, 'Letter no. 13089', https://www.darwinproject.ac.uk/letter/?docId=letters/DCP-LETT-13089.xml

23. Letter to B.J. Sulivan, 20 Mar. 1881, Darwin Correspondence

Project, 'Letter no. 13092', https://www.darwinproject.ac.uk/letter/?docId=letters/DCP-LETT-13092.xml

24. Letter to B.J. Sulivan, 1 Dec. 1881, Darwin Correspondence Project, 'Letter no. 13525', https://www.darwinproject.ac.uk/letter/?docId=letters/DCP-LETT-13525.xml

25. See 'Darwin's Missionary Endeavour', *Church Times*, 13 Feb. 2009.

26. Cited in *Witnessing under the Southern Cross: Mr. and Mrs. Burleigh at Wollaston and Tekenika* (London: South American Mission Society, 1902), https://anglicanhistory.org/sa/burleigh1902.html

18. THE GENOCIDE OF THE SELK'NAM

1. Carlos Gallardo, *Tierra del Fuego: Los Ona* (Buenos Aires: Cabaut, 1910),109–10.

2. Documentary, *La Patagonia en el año 1910*, Museo Maritimo y del Presideio de Ushusia, YouTube, 20 May 2020, https://youtu.be/copSpkomriU?si=YdyQ0a58ox2Vmjrg

3. Quoted in Albert Harambour, '"There Cannot Be Civilisation and Barbarism on the Island": Civilian-Driven Violence and the Genocide of the Selk'nam People of Tierra del Fuego', in Mohamed Adhikari (ed.), *Civilian-Driven Violence and the Genocide of Indigenous Peoples in Settler Societies* (London: Routledge, 2019), https://www.researchgate.net/publication/352158823_'There_Cannot_be_Civilisation_and_Barbarism_on_the_Island'_Civilian-driven_Violence_and_the_Genocide_of_the_Selk'nam_People_of_Tierra_del_Fuego1

4. Ramon Lista, *Viaje al País de los Onas: Tierra del Fuego* (Buenos Aires: A. Nuñez, 1887), 23.

5. Spears, *Gold Diggings of Cape Horn*, 6.

6. For a description of this encounter, see Julio Popper, *Exploration of Tierra del Fuego: A Lecture Delivered at the Argentine Geographical Institute on the 5th of March, 1887* (Buenos Aires: L. Jacobsen, 1887), 80.

7. Duncan S. Campbell and Gladys Grace-Paz (eds), *Patagonia Wild and Free: Memoirs of William Greenwood*, digital edn (n.p.: Duncan S. Campbell and Gladys Grace-Paz, 2017), 155.

8. For a granular history of the Menéndez/Braun commercial dynasty in its formative years in Punta Arenas, see Marchante, *Menéndez, rey de la Patagonia*.

9. Harambour, '"There Cannot Be Civilization"', 175.

10. 'Patagonia: A New Zealander's Impressions', Patagonia Bookshelf. From Duncan S. Campbell and Gladys Grace Paz website: https://patlibros.org/nzl/ltr2.php?lan=eng

11. See Campbell and Grace-Paz, *Patagonia Wild and Free*, 98–9.

12. Bridges, *Uttermost Part of the Earth*, 349.

13. Ibid., 371.

14. Quoted in Carlos Gigoux, '"Condemned to Disappear": Indigenous Genocide in Tierra del Fuego', *Journal of Genocide Research*, vol. 24, no. 1 (2022), 1–22.

15. Letter in *Sumario sobre vejámenes inferidos a indíjenas de Tierra del Fuego*, Punta Arenas, Legajo 75, Archivo Judicial de Magallanes, 2 Dec. 1895.

16. Case details in *Sumario sobre vejámenes*.

17. For details of Lehmann-Nitsche's expeditions to Tierra del Fuego, see Diego Ballestero, 'Picturesque Savagery on Display: Exhibition of Indigenous People, Science and Commerce in Argentina (1898– 1904)', *Anthropological Journal of European Cultures*, vol. 31, no. 2 (2022), 65–88. Also, Inés Yujnovsky, 'Viajeros a la sombra de Darwin en los confines del siglo XIX argentino' (PhD diss., El Colegio de México, Centro de Esudios Históricos, 2010), https:// repositorio.colmex.mx/concern/theses/dz010q37g?f%5Bcenter_ sim%5D%5B%5D=Centro+de+Estudios+Históricos&f%5Bcreator_ sim%5D%5B%5D=Yujnovsky%2C+Inés&locale=es

18. Pascal Blanchard et al. (eds), *Human Zoos: Science and Spectacle in the Age of Colonial Empires* (Chicago: Chicago University Press, 2009), 144.

19. Marchante, *Menéndez, rey de la Patagonia*, 50.

20. Marchante, *Menéndez, rey de la Patagonia*, 214–15.

21. José María Beauvoir, *Aborígenes de la Patagonia: Los Onas, tradiciones, costumbres y lengua* (Buenos Aires: n.p., 1915), 27–8.

22. Armando Braun Menéndez, *Pequeña historia fueguina* (Buenos Aires: n.p., 1946), 71.

23. H.G. Wells, *The War of the Worlds* (London: Plain Label Books, 1967), 6.

24. Ministerio del Interior de Chile, 'Informe de la Comisión Verdad Histórica y Nuevo Trato con los Pueblos Indígenas' (2003), 478 https://bibliotecadigital.indh.cl/bitstreams/14dcc8ee-3740-46ff- b520-a34ae85c12ca/download

25. 'Chile Repatriates Tribal Remains from Zurich', Swissinfo.ch, https:// www.swissinfo.ch/eng/culture/chile-repatriates-tribal-remains- from-zurich/8098352

19. THE WILD SOUTH

1. Alberto B. Martínez, *Baedeker of the Argentine Republic* (New York: D. Appleton & Co., 1916), 56–7. This was not a genuine version of the famous travel guide but a scrupulously researched 'fake Baedeker', written by an Argentinian named Alberto Martínez, intended to capitalize on Argentina's new international status.

2. For an analysis of these different scientific and intellectual responses, see Julia Rodriguez, *Civilizing Argentina: Science, Medicine, and the Modern State* (Chapel Hill, NC: University of North Carolina Press, 2006).

3. 'Censo de población de los territorios nacionales, República Argentina' (Buenos Aires, 1914), https://biblioteca.indec.gob.ar/bases/minde/1c1912_1.pdf

4. 'Censo de la República de Chile: Levantado el 28 de noviembre de 1907', https://www.memoriachilena.gob.cl/602/w3-article-92761.html

5. Bailey Willis, *A Yanqui in Patagonia* (Stanford: Stanford University Press, 1948), 97–8.

6. For a study of the ideas and practices of the penal colony 'panopticon', see Ryan Edwards, 'From the Depths of Patagonia: The Ushuaia Penal Colony and the Nature of the "End of the World"', *Hispanic American Historical Review*, vol. 94, no. 2 (2014), 271–302.

7. Alberto Martínez and Maurice Lewandowski, *The Argentine in the Twentieth Century* (Buenos Aires: n.p., 1915), 163.

8. W.A. Hirst, *Argentina* (London: T.F. Unwin, 1910), 288.

9. Ramón Lista, *Los Indios Tehuelches: Una raza que desaparece* (Buenos Aires: Imprenta de Pablo É. Coni, 1894), 12.

10. H. Hesketh Prichard, *Through the Heart of Patagonia* (New York: D. Appleton and Company, 1902), 101.

11. Colonel Sir Thomas Hungerford Holdich, *The Countries of the King's Award* (London: Hurst and Blackett, 1904), 233–4.

12. Roberto Payró, *La Australia Argentina* (Buenos Aires: Imprenta de La Nación, 1898), 232.

13. Ibid., 81.

14. *Censo de la República de Chile: levantado el 28 de noviembre de 1907*: Censo de la República de Chile : levantado el 28 de noviembre de 1907 - Memoria Chilena, Biblioteca Nacional de Chile.

20. SALVAGE AND SURVIVAL

1. Karl Pearson, *National Life from the Standpoint of Science* (London: A. and C. Black, 1905).

2. Quoted in Daniel Tumarkin, 'Miklouho-Maclay: 19th Century Russian Anthropologist and Humanist', *RAIN*, vol. 51 (Aug. 1982), 4–17. See also Goergy S. Levi and Uwe Hossfeld, 'Ernst Haeckel, Nikolai Miklucho-Maclay and the Racial Controversy over the Papuans', *Frontiers of Zoology*, vol. 17, no. 16 (2020).

3. Anténor Firmin, *The Equality of the Human Races* (Urbana: University of Illinois Press, 2002), 450.

4. Franz Boas, *The Mind of Primitive Man* (New York: Macmillan, 1921), 278.

5. See the *UNESCO Courier: A Window Open on the World*, nos. 8–9, https://unesdoc.unesco.org/ark:/48223/pf0000070030

6. Quoted in Julius E. Lips, *The Savage Hits Back* (New York: University Books, 1966), vii.

7. Charles Wellington Furlong, 'The Vanishing People of the Land of Fire', *Harper's Magazine*, Jan. 1910.

8. I.H. Coriat, 'Psychoneuroses among Primitive Tribes', *The Journal of Abnormal Psychology*, vol. 10, no. 3 (1915), 201–8.

9. Ricardo Rojas, *Archipiélago: Tierra del Fuego* (Ushuaia: Editorial Sudpol, 2014), 47–9.

10. Potenze, *Científicos y religiosos*, 193–4.

11. Anne Chapman, *End of a World: The Selknam of Tierra del Fuego* (Buenos Aires: Zagier & Urruty Publications, 2008), 13–14.

12. Anne Chapman, *European Encounters with the Yamana People of Cape Horn, before and after Darwin* (Cambridge: Cambridge University Press, 2010), 2.

13. For example, 'Chilean Indigenous Language Vanishes as Last Living Yamana Speaker Dies', *The Guardian*, 17 Feb. 2022.

14. 'Copacho, último Tehuelche de la Patagonia austral', *Caras y Caretas*, 1939, Archivo General de la Nación Argentina. AGN-AGAS01-rg-Caja 2508-Inv:153248.

15. 'Muere en Río Gallegos Capipe, el último de los tehuelches', *El Diario*, 13 July 1953.

16. 'Muere en Río Gallegos Capipe, el último de los tehuelches', *El Diario*, 13 July 1953. 'Remembering Dora Manchado, the Last Speaker of Tehuelche', Living Tongues Institute, 7 Jan. 2019, https://livingtongues.org/remembering-dora-manchado/

17. 'Censo de población y Vivienda', Instituto Nacional de Estadísticas, Chile, 2017, https://redatam-ine.ine.cl/redbin/RpWebEngine.exe/Portal?BASE=CENSO_2017&lang=esp

18. 'Censo 2022', https://censo.gob.ar/#:~:text=%C2%BFCu%C3%A1ntos%20somos%20en%20la%20Argentina,Argentina%20somos%2046.044.703%20habitantes

19. Pueblos Originarios, '"Soy de raíces indígenas": Campaña de cara al próximo censo nacional', Resumen Latinoamericanco, 14 Mar. 2022.

20. Rodolfo Stavenhagen, 'The Return of the Native: The Indigenous Challenge in Latin America', University of London, Institute of Latin American Studies, Occasional Papers no. 27 (2001).

21. 'The Museum of Patagonia Opens Its Doors', Benetton Group, press release, 12 May 2000, https://www.benettongroup.com/en/media-press/press-releases-and-statements/the-museum-of-patagonia-opens-its-doors/

22. 'UN Declaration on the Rights of Indigenous Peoples', Office of the High Commissioner for Human Rights, 13 Sept. 2007, https://www. ohchr.org/en/indigenous-peoples/un-declaration-rights-indigenous-peoples

EPILOGUE: THE HOUSE OF THE WIND

1. 'The Yagan's Letter to the King of Norway: "Don't Install This Destructive Industry in Our Territory"', *Patagon Journal*, 5 Apr. 2019, https://www.patagonjournal.com/index.php?option=com_content &view=article&id=4207%3Ala-carta-de-la-comunidad-yagan-a-los-reyes-de-noruega-no-instalen-esta-industria-destructiva-en-nuestro-territorio&catid=60%3Aeditor&Itemid=264&lang=en

2. See Gustavo Blanco-Wells et al., 'Plagues, Past, and Futures for the Yagan Canoe People of Cape Horn, Southern Chile', *Maritime Studies*, vol. 20 (2021), 101–13, https://doi.org/10.1007/s40152-021-00217-2

3. 'Everyone Equal: The Resilience of Indigenous Peoples across the Globe', World Bank Group, 7 Aug. 2020, https://www.worldbank. org/en/news/feature/2020/08/07/everyone-equal-the-resilience-of-indigenous-peoples-across-the-globe

4. For an analysis of these expectations, see Chapter 5, 'Dreamworlds of Beavers', in Laura A. Ogden, *Loss and Wonder at the World's End* (Durham, NC: Duke University Press, 2021). See also Mara Dicenta, 'White Animals: Racializing Sheep and Beavers in the Argentinian Tierra del Fuego', *Latin American and Caribbean Ethnic Studies*, vol. 18, no. 2 (2023), 308–29. Available at: https://www.tandfonline.com/doi/full/10.1080/17442222.2021.2015140#abstract

5. Katie Gopal, 'Indigenous Group Asks SEC to Scrutinize Fracking Companies Operating in Argentina', Covering Daily, 30 Sept. 2024, https://www.coveringdaily.com/climate/82176.html

6. 'An Imported Tree Fuels Patagonia's Terrifying Summer Fires', *National Geographic*, 4 Jan. 2022.

7. Mario Vargas Llosa, 'Questions of Conquest: What Columbus Wrought, and What He Did Not', *Harper's Magazine*, Dec. 1990.

8. 'Call for Napalm Bombing of "Savages" Wins Survival Racism Award', 26 Aug. 2009, Survival International, https://www.survivalinter national.org/news/4885

9. See https://www.survivalinternational.org/stampitout

10. Author's translation from Niní Bernadello (ed.), *Cantando en la Casa del Viento: Poetas de Tierra del Fuego* (Ushuaia: Editoria Cultural Tierra del Fuego, 2020), 20.

ACKNOWLEDGEMENTS

This book could not have been written without the scholars, historians and activists who came to this material long before I did. I walk, to some extent, in their footsteps, and any errors or inaccuracies are mine alone. Thanks also to everyone who agreed to be interviewed: Marcelo Valko, Victor Vargas Filgueira, Richard Yasic Israel, Analía Lanteri, Gustavo Barrientos, Máximo Farro, Maurice van de Maele and Fernando Pepe Tessaro and Marcos Bufándo Fernández at the Colectivo GUIAS. Thanks to the Museo Roca, the Archivo General de la Nación (Argentina) in Buenos Aires, Franklin Pardon Cárdenas at the Museo Regional de Magallanes in Punta Arenas, the Museo de la Patagonia Francisco P. Moreno in Bariloche, to Don Jorge and the staff at the Casa Ronco in Azul, and the Museo de La Plata. A generous grant from the Society of Authors went a long way to financing my travels to the far south. I owe a special debt of gratitude to my publisher Michael Dwyer, for lending a helping hand at a difficult moment. Thanks, as always, to my wife Jane, who was with me throughout my Patagonian travels, as she has been ever since the longer journey we began together thirty-four years ago.

INDEX

INDEX

INDEX

INDEX